THE SENSE OF TIME

THE SENSE OF TIME

An Electrophysiological Study
of Its Mechanisms in Man

Josef Holubář

Translated from the Czech
by John S. Barlow

THE M.I.T. PRESS
Cambridge, Massachusetts, and London, England

Originally published in Czech in 1961
by State Medical Publishing House, Prague,
under the title "Časový Smysl."

Copyright © 1969 by
The Massachusetts Institute of Technology

Set in Monotype Century Expanded.
Printed and bound in the
United States of America by The Riverside Press.

All rights reserved. No part of this book may be reproduced in any form or by any means, electronic or mechanical, including photocopying, recording, or by any information storage and retrieval system, without permission in writing from the publisher.

SBN 262 08034 6

Library of Congress catalog card number: 76-87294

FOREWORD

Despite the widespread general interest in the psychological and physiological aspects of the sense of time, there have been relatively few direct experimental explorations of its fundamental mechanisms in the brain. The possibility that rhythms in the electroencephalogram (EEG) of man, in particular the alpha rhythm of some 10 cycles per second, might in some way reflect fundamental timing mechanisms in the brain, at least for relatively brief periods, has occurred to a number of workers over the years. The early findings of Hoagland, that an increase in body temperature was accompanied by an underestimation of time intervals and also by an increase in the alpha frequency (although the two differed with respect to their temperature characteristics), and of Jasper and Shagass, that temporally conditioned alpha-blocking was a more accurate measure of time intervals than were the subjects' own estimates, lent some support to such a notion, as has more recent work by other authors.

But some writers have gone further, proposing that the alpha rhythm itself is a specific manifestation of a rhythm for the measurement of time. Norbert Wiener was a strong proponent of this concept, and it was also with this idea that Dr. Holubář began his own work. His particular contribution, however, lies in using as his experimental tool the well-known effect of flicker or photic stimulation on the EEG (i.e., photic driving or photic entrainment), in combination with temporal conditioning (or, conditioned reflexes to time, as it is expressed in Czech, and in Russian). His remarkable findings, that the intervals of temporally conditioned galvanic skin responses can be specifically altered by flicker in a manner that is determined by the relation between the rates of flicker and the frequency of the alpha rhythm, have however largely escaped attention, perhaps because his publications on this work in other than the Czech language have either been relatively inaccessible, or have appeared only in the form of brief abstracts.

It was late in 1960 that I first learned of Holubář's work from Professor Herbert Jasper, and shortly thereafter Professor Norbert Wiener (with whom I shared a deep interest in the problem of "brain clocks") showed me a copy of Holubář's monograph, which had then just been published. For several reasons, the work, as described in the English summary, was immediately of interest to me, but it was not until recently that, having gained some knowledge of the Czech language, I was able to read the monograph in its entirety, and came to the conclusion that it merited the wider attention that would be possible with a translation into English. For the monograph not only includes a full account of Holubář's own experimental work, it also includes a rather unique survey of the literature (including the Russian) which, although by no means complete, as he himself points out, does draw from an unusually wide range of fields in relation to the general problem of the sense of time.

The fact that the monograph was originally published in 1961 of course necessarily precludes mention of more recent work in the areas that Dr. Holubář surveyed, as well as current views on such problems as the physiological basis of the EEG, and the navigation by birds, to mention only two topics that are included in his discussion. It seemed advisable then to compile a supplementary bibliography to accompany the present translation which would reflect, although again with no intent of completeness, some of these more recent developments. Holubář's own experimental work, however, remains a stimulating and challenging contribution.

Although his work in recent years at the Institute of Physiology of the Czechoslovak Academy of Sciences in Prague had been primarily in other areas of electrophysiology, Holubář maintained a continuing interest in the sense of time, and he and I had begun to make some preliminary plans for a collaborative endeavor on some further aspects of the problem in relation to flicker and the EEG in man, when the tragic news arrived from Prague in October 1967 that he had been fatally injured in a traffic accident. The present translation is then dedicated to the memory of this kind and gentle man. Perhaps with the wider availability of his work, further steps—some that he himself suggested in his concluding remarks—may be taken which will help to unravel the enigma that is the physiological basis of the sense of time. (In the field of bird navigation, a phenomenon in which physiological chronometry is generally agreed to be of first importance, such a contribution would be especially appropriate, for as he mentioned to me during our first meeting in Prague in 1961, the word "holubář" in Czech means "pigeon breeder.")

I am very grateful to Dr. Vilém Kuthán, of the Institute of Physiology of the Czechoslovak Academy of Sciences in Prague and a colleague of Dr. Holubář, for reviewing the translation, which was arranged at my request by Dr. Holubář's father, Professor Josef Holubář. Professor Holubář also very kindly provided a copy of the original dissertation from which the illustrations for the present translation were reproduced.

JOHN S. BARLOW, M.D.

Department of Neurology
Massachusetts General Hospital and
Harvard University Medical School
Boston, Massachusetts
August 21, 1968

PREFACE

The thought of investigating the mechanism of the sense of time by means of interrupted light, which is developed and realized in this work, occurred to me in the year 1953. The first preliminary experiments were carried out in the following year, and in successive years some preparatory experiments were undertaken, entirely unsystematically, in the area of other work. Only at the beginning of the year 1957 was systematic work begun on the main series of experiments, with extensive electroencephalography.

Since a separate, long-term, and possibly unproductive investigation could not be risked, several subordinate questions that were marginal to the fundamental problem were pursued in the course of the main experimental series; these led to some findings that are mentioned in this work but are not essential to the main idea.

In addition to the preliminary publications mentioned in the references, the opportunity occurred a number of times of presenting the results of this work in lectures to the scientific public. Besides individual communications to the physiological meetings in Prague in January 1958, in Hradec Králové in June 1958, and in Bratislava in June 1959, the entire subject matter not concerned directly with the sense of time was presented to the Czechoslovak-Polish EEG symposium in Cracow in June 1958, and the principal theme on the sense of time was presented to the Third Physiological Meeting in Brno in January 1959, to the International EEG Symposium in Jena in October 1959, to the working conference of the Commission of Electroneurologists in Prague in November 1959, to the First Czechoslovak Conference on Rocket Technology and Astronautics in Liblice in April 1960, and to a seminar at the Psychiatric Clinic in Prague in November 1960. In addition, I had the opportunity of discussing the principal findings at the International Symposium on Reflex Mechanisms in the Pathogenesis of Epilepsy in Liblice in September 1960.

Prague, January 1961
J. HOLUBÁŘ

CONTENTS

Foreword   v

Preface   ix

Introduction   1

The Question of the Existence of a Sense of Time   3

    Temporally Conditioned Reflexes   6
    The Navigation of Birds   10

Brief Review of the Psychological Literature on the Sense of Time   14

    Methods of Investigation   15
    Principal Concepts   16
    The Significance of Different Factors   18
    The Basis of the Sense of Time   22

The Working Hypothesis and Plan of Its Experimental Verification   26

    Biological Rhythms   27
    Brain Rhythms   29
    Plan of the Experiments   31

Methods 33

    Experimental Subjects 33
    Apparatus and Experimental Procedures 34

Results and Their Analysis 39

    Preliminary Series 39
    Galvanic Skin Reflex (with the Aid of an Auxiliary Current) and Its Conditioning 42
    Trace-Conditioned Galvanic Skin Reflex 44
    Main Experimental Series 45
    EEG Changes Accompanying the Galvanic Skin Reflex 47
    Dissociation of the Galvanic Skin Reflex and the Accompanying EEG Changes 59
    Temporally Conditioned Reflexes 62
    The Specific Effect of Flicker on the Intervals of Temporally Conditioned Reflexes 69
    Discussion of the Results from the Standpoint of the Working Hypothesis 74

Conclusion 82

References 86

Supplementary Bibliography 99

Epilogue: In Memoriam 109

Index 111

THE SENSE OF TIME

INTRODUCTION

> *In the course of 42 years of autonomous time that is measured on board a rocket (granting an enormous velocity, commensurate with the velocity of light), it would be possible to fly around the entire known universe. For time as measured on the earth, however, the corresponding lapse would be approximately three billion years. Upon the return of the rocket, neither the earth nor the solar system would still exist.*
> (E. Sänger, Freudenstadt, 1956)

If today, on the threshold of the space age, physicists promise us such unbelievable possibilities as matter-of-course consequences of Einstein's theories, which have been constantly confirmed by experience, and if from such a brilliant theory has gradually emerged the physical basis of a world-known view of modern man, certainly it is also appropriate for the sense of time to become anew a subject of biological investigation. By the sense of time is meant the ability of man and other living organisms to perceive the flow of time, to act in specific temporal patterns, as for example in specific periodic rhythms, and also to appreciate time and the temporal aspects of all sensory perceptions, simply with the aid of intrinsic means of the organism for measuring time.[1] Gellershtein (1958) demonstrated the considerable practical significance of the study of the sense of time for physical culture and numerous other human activities.

Time is such a common attribute of all natural phenomena that for us the temporal aspect of sensory perception and other processes of life seems self-evident, as a rule. The significance of the time dimension is prominent by comparison (Davis 1956); whereas, for example, in visual perception the spatial aspect clearly predominates, in auditory perceptions it is the temporal aspect i.e., the rhythm, temporal sequence, or the duration of sensations, that clearly predominates. The organism has the ability to estimate quantitatively the duration of events perceived, to remember the temporal sequence of events, and to produce—comparatively very accurately—specific time intervals. These are undeniable facts, which will be documented here by numerous examples.

---

[1] In order to avoid error, it should be mentioned that the dependence of time on the velocity of motion is purely a physical phenomenon which does not have anything in common with "biological time" or with the time sense of living beings; it is only with this reservation that the quotation from Sänger was used in the Introduction (see also Dvořák 1959). Life, to be sure, necessarily runs its course in physical time, as in physical space. Acceleration has powerful biological effects, but velocity does not (Čapek 1953).

THE QUESTION OF THE EXISTENCE OF A SENSE OF TIME

At the outset it is necessary to settle the question: is it at all correct to presume some kind of sense of time, some kind of biological mechanism, by which organisms measure time? I shall leave aside von Skramlik's (1939) criticism of the concept of the sense of time, which is rather semantic in nature (he concludes that such a concept is unwarranted for the reason that no organ for the sense of time is known, whereas the organs for the senses of vision, hearing, etc., are common knowledge). Woodrow (1951) surveyed the various theories on the perception of time and also the concept that the duration of perceptions is never perceived directly and independently but only as an attribute of phenomena which invariably presuppose some kind of process of change. If there is no change, then time is at a standstill. The perception of time is only continuous afferentation, brought about by changes in the environment. A further analogy could be suggested: there is no special sense for time, the same as there is no special sense for space.

This view, however, is incorrect. To start with, let us examine the above-mentioned analogy: there exist several perfected mechanisms for the perception of space that are the same in principle for different senses, that is, the projection of receptors to various levels of the central nervous system, especially to the cerebral cortex, such as the cortical projection of the retina and the cortical projection of somatic sensation. It is a fundamental principle, even though simple and self-evident, that a necessary condition for visual or tactile perception, for example, is continuous spatial systematization or organization in the course of perception. That this is not the only principle possible, we see for example from the transmission of an image by television, which is created entirely differently— by a temporal resolution of the image, i.e., by the temporal sequence of signals and an exact synchronization of the receiver with the transmitter.

Further, it must be asked, if the ability of an organism to measure time is actually demonstrated, is it not actually a fallacy, in the sense that time is measured by means of a clock or some other physical apparatus from which the organism merely reads time? This objection must be taken very seriously; it is far more important than perhaps appears initially; nevertheless, it can be contradicted.

Estimates of time can be very inaccurate, depending on the purpose of the experience, especially during an emotionally colored experience. From complementary comparison with objective time, such sensations arise as "time flowing rapidly," or on the other hand "lengthy boredom." Here, to be sure, other factors interfere, e.g., interest or emotion, so that the sense of time must be studied under conditions in which other factors are excluded as far as possible, especially in cases when the process of estimation of time is subconscious. Even here, however, it is often possible to show considerable accuracy of time judgment in individuals with extensive experience. From everyday experience we know very well how long the usual daily acts, such as crossing a room or a street, reading pages of a book, writing lines of a text, and so forth last approximately; this enables us to estimate with sufficient accuracy—even subconsciously—the duration of any sort of event composed of such elements.

Frequent mention is made in reports of the existence of the sense of time in so-called posthypnotic suggestion, but this is entirely incorrect. Subjects, in response to a suggestion given them under hypnosis, carry out some simple act (such as, opening a window) upon awakening from hypnosis, perhaps the next day, at a specified hour, at the exact time specified under hypnosis. In this connection, it is remarkable that the subject carries out the task without being aware that it had been carried out as a kind of order, and that he often subsequently looks for a reason for his action. There is here, however, no peculiarity in that the task was carried out at a specified time, since there exist many other acts usually carried out consciously at a specific time, and there exist a limitless number of events in our environment on which man can rely for time estimation on the basis of experience.

Similarly, the existence of a time-measuring mechanism in the organism is not proved by the frequently mentioned ability of some people to awaken exactly at a time predetermined before falling asleep. Such a biological mechanism presupposes in the first place the existence of some kind of internal rhythm which could serve like the pendulum of a clock for the counting out of time intervals and thus for the measurement of time. However, under rigorous experimental conditions, as for example in the work of Omwake and Loranz (1933), in which the ability of people to waken themselves at a predetermined time was unquestionably demonstrated, a clock was available on the bedside table of the experimental subjects, and hence the time-measuring mechanism was in the environment, not in the body.

I assume that there are two groups of phenomena that demonstrate beyond all doubt the existence of a special sense of time, that is, a mechanism by which organisms can measure time: There are on the one hand temporally conditioned reflexes (temporal conditioning), and on the other hand the navigation of birds, or experiments which support a theory of navigation. Both groups of phenomena are practically inexplicable without the presumption of the existence of a time-measuring mechanism in the organism; from such a presumption a clear picture of a large collection of facts can be given without substantial contradiction.

Temporally Conditioned Reflexes

These were discovered in Pavlov's laboratory. Zeleniĭ (1907) noted that the reinforcement of the salivary conditioned reflex in dogs to sound at regular intervals of 10 minutes resulted subsequently in salivation in the same regular intervals but without the presentation of any stimuli. Krzhishkovskiĭ (1908), during a study of inhibition, established that a conditioning stimulus was always effective between 33 and 34 minutes but was ineffective between 19 and 20 minutes. He then began purposefully to establish a conditioned salivary reflex by reinforcement each 10 to 13 minutes, and thus elaborated a temporally conditioned reflex.

The most notable of the early objective work on temporal conditioning was that of Feokritova (1912). Conditioned salivary reflexes were established in three dogs to the sound of a metronome, with reinforcement at regular intervals of 10, 15, or 30 minutes, respectively. Altogether, 200 or more reinforcements were presented; in some cases other stimuli (the noise of a whistle, air, a phonograph) were also employed. It was shown that an important attribute of temporally conditioned reflexes is the great accuracy that they manifest after they have become established. By means of the method of differentiation, an accuracy of as great as one minute, i.e., ±3 per cent, was achieved. Further, Feokritova ascertained that the extinguishing of temporally conditioned reflexes does not occur gradually, as for the usual conditioned reflexes; instead, they become extinguished at once, similar to trace-conditioned reflexes (Pimenov 1907, Zavadskiĭ 1908, Grossman 1909, Dobrovolskiĭ 1911).

Stukova (1914) continued the study of temporally conditioned reflexes, especially investigating the inhibitory influences of the environment on them. Strong faradic stimuli applied at times when a temporally conditioned reflex should occur completely inhibited the reflex, so that it did not appear. She further studied the impairment of the rhythm of temporally conditioned reflexes that resulted from internally administered caffeine and cocaine.

The problem of temporal conditioning was further investigated by Deriabin (1916), who established that the nature of conditioning stimuli applied during the establishment of temporally conditioned reflexes is immaterial; what is essential is the time interval for the stimuli. In conformity with this, he established that dogs learn to differentiate a stimulus according to its rhythm, without regard to the kind of sound, i.e., that they "abstract the time factor" (for example, differentiating an auditory stimulus whose characteristic is invariable alternation of a 2-second sound and a 2-second silence, whether the sound is from a whistle, a trumpet, or otherwise).

The significance and appearance of temporally conditioned reflexes during dynamic stereotypes (patterned movements) were studied, for example, by Vasilenko

(1932), Maiorov (1933), and Vatsuro (1948). Defensive movement as a temporally conditioned reflex has been described by Beritov (1932). Later they were studied by numerous other authors (Gambarian 1952, Bolotina 1952a, b, 1953, Dmitriev and Kochigina 1955), generally for intervals of 3 to 5 minutes, only exceptionally (under the influence of pharmacological agents) for only 1 minute. Under no conditions was it possible to shorten the interval further. The great accuracy of temporally conditioned reflexes was confirmed. It was shown that for the appearance and accuracy of a conditioned reflex the use of a specific conditioning stimulus is unnecessary, although the latter somewhat facilitates the establishment of the reflex.

Temporal conditioning has been described in the turtle (Nikiforovskiĭ 1929), in the pigeon (Baiandurov 1937), in the monkey (Bolotina 1952b), and in many other animal species (Nikiforovskiĭ 1951). The regularity is in essence the same in every instance, taking into account of course the particularities of the individual species. In animal behavior, many natural temporally conditioned reflexes can be seen, in particular the 24-hour rhythms of numerous functions (see Dmitriev and Kochigina 1955). Kvasnitskiĭ and Koniukhova (1954) directed attention to the practical importance of knowledge of such natural conditional reflexes to time in the husbandry of cattle. Delaying the milking time by 30 to 40 minutes results in a decrease in the quantity by 5 per cent, as well as a decrease in the quality.

In man, temporally conditioned reflexes have not been studied as extensively. Bekhterev (1908) described the reflex in man, during the establishment of a rhythmic stereotype. Motor responses to a rhythmically repeating sound continued even after cessation of the stimuli—the more prolonged the repetition, the stronger the reinforcement. Frolov (1951) stressed a tendency toward a small overestimation of the time interval during experiments carried out in silence and darkness. The method adopted by Zeleniĭ (1923), in which the subject claps his hands or taps his feet according to the sound of a metronome, so that temporal conditioning appears as a continuation of the clapping during silence of the metronome, has been repeatedly employed (for example by Alekseev 1953). Dmitriev and Kochigina (1955) studied temporal conditioning established by verbal reinforcement at intervals of 25 to 30 sec, in children 8 to 14 years old. Hull (1943) established temporally conditioned galvanic skin reflexes with intervals of 30 sec in man.

Kanaev (1956), using the method of key-pressing at intervals of 3, 6, or 12 minutes, respectively, achieved the extraordinary accuracy of 1 to 2 seconds (i.e., approximately 0.3 per cent) in children 9 to 14 years of age, who were rewarded with candy.

Jasper and Shagass (1941a, b) conditioned the light-induced blocking of the alpha rhythm by means of sound stimuli and also established other temporally conditioned reflexes. The intervals of the former after establishment were more exact than those of attempted conscious judgment (signaled by pressing a key) which was carried out at the same time. The authors concluded that the mechanisms of the two manifestations of the sense of time would perhaps be different.

There have been two interpretations of the mechanism of temporally conditioned reflexes: Pavlov pointed to the large number of different rhythmic phenomena constantly occurring in the organism and concluded that during the establishment of temporally conditioned reflexes there arose a connection between some suitable internal rhythm and the rhythm of the reflex established; i.e., rhythms of repetitive processes in the organism in fact become the conditional stimuli for the newly established conditioned reflex. Hull (1943) introduced another possible explanation: By constant repetition of a reflex, a situation is created such that the end of the course of the response or a certain phase of a late consequence of the response provides, as a consequence of its temporal position immediately before the succeeding stimulus and response, a conditioned stimulus for the repetitive response.

From what has been mentioned concerning temporally conditioned reflexes and experiments on their interpretation, I believe the conclusion very clearly follows that human and animal organisms unquestionably have at their disposal a time-measuring mechanism and that the existence of the sense of time in living beings is well founded.

The Navigation of Birds

The extraordinary orienting ability of migratory birds and homing pigeons, which was already well known in ancient times, has been confirmed by modern science; such birds can find their destination over a distance of many hundreds or thousands of miles under conditions where the method of orientation seems entirely enigmatic. Birds, for example, depart immediately in the correct direction from places where they had never been previously and to which they had been transported without any possibility of sensory orientation.

In contrast to a long series of the most varied, frequently fanciful, and in the end always unsuccessful explanations that had been offered, it was only in recent years (Matthews 1955) that a complete theory has been offered, which includes all the pertinent facts, is very convincingly supported by experiment, and seems at last to resolve fundamentally the problem of orientation of birds during long flights, and also that of bees (von Frisch and Lindauer 1954).

According to this theory, birds carry out their navigation in relation to the sun. The arc which the sun circumscribes in its diurnal orbit indicates, as is well known, by means of its inclination with respect to the horizon (making allowance for the time of year), the latitude and, by means of the time when the sun attains the pinnacle of the curve or when noon occurs, the geographical longitude (each 15 degrees in an easterly direction is equivalent to the meridian of one hour earlier). This is the basis of determination of geographical position on the earth and of navigation by mariners. Birds of course do not measure the elevation of the sun and do not carry out the necessary calculations as does man. In further commentary we shall limit ourselves to the navigation of homing pigeons.

Pigeons have a knowledge of the course of the sun's arc at the location of their home, a day-to-day correction being made from memory, as the elevation of the sun above the horizon changes according to the time of the year. If the bird finds itself in another place, it notes the change of the sun's arc in comparison with its home. It must be presumed that in the course of a short-time "search for direction," the pigeon carries out an extrapolation of the sun's arc. If the arc is found to be too low (as compared with home), the bird flies toward the arc (i.e., toward the south) during which the arc rises and approaches the form of the home arc. On the other hand, if the sun's arc is too high, the pigeon corrects the height of the latter by flight from the arc (i.e., toward the north). In addition, however, the pigeon notes, thanks to its sense of time, its "chronometer," which constantly gives the local time of its home, that the sun on its diurnal course is either delayed or advanced. In the first instance, the bird flies against the sun (i.e., toward the east), thereby accelerating the relative motion of the sun somewhat; in the second case, the bird flies after the sun (i.e., toward the west), as though to overtake it. Thus in both cases the bird adjusts the appearance of the sun's arc in such a way as to resemble the home arc more closely, and in this way to approach home. The navigation is sufficiently accurate to guide the bird close enough to home that the latter can then be found from the local characteristics of the area.

A necessary presumption, therefore, in the navigation of birds is the existence of an accurate sense of time which functions—at least for a certain period—independently of the environment, so that the bird can detect any change in the environment with respect to its internal time. This is most important for our requirements, but the experimental reports to which I shall refer later attest that it is real and that birds therefore indeed have such a sense of time at their disposal.

Matthews (1955) cites a series of reports in favor of this theory of navigation. A superiorly developed labyrinth enables the pigeon to maintain its head constantly in a horizontal position even during flight and thereby to estimate the altitude of the sun precisely. The highly developed vision of birds is well known; pigeons have a visual acuity three times sharper than man, and in addition this high acuity encompasses the whole visual field, not only central vision. Therefore the ability for perceiving the movement of objects in the visual field is also exceptionally superior. All of these attributes of the senses enable the extrapolation of the sun's arc with sufficient accuracy for navigation. Such navigation applies for distances greater than approximately 30 to 50 miles, at which point the difference in the sun's arc becomes sufficiently large; for shorter distances an initially correct direction of flight is impossible, and hence to a large degree a random course must be used.

The necessity of the sun for the orientation of pigeons has been demonstrated experimentally beyond all doubt. In the presence of clouds, orientation is seriously impaired, and under a complete overcast or at night it is completely incapacitated. Actually, visibility of the sun even for only a short time suffices for the assumption of the direction, which is then maintained with small errors and for a long time without the sun. Under laboratory conditions, birds can be trained to a specific direction given by an artificial sun or by the natural sun altered with the aid of a mirror.

In these experiments, however, it is also of major importance that the birds learn a specific direction with reference to the sun at a specific time of day; at any other time, the birds change the angle with respect to the sun in such a way that they maintain the same direction with respect to the earth. The birds must therefore have a knowledge of time independently of the sun; they must therefore possess their own "chronometer." This is indicated particularly persuasively by experiments in which the internal time was artificially altered. Hoffman (1953) trained starlings to feed at a place oriented in a specific direction with respect to the starting point. Then the birds were kept for 12 to 18 days under artificial lighting conditions which delayed their time by 6 hours. In the decisive experiment, the orientation of the starlings was shifted by 90 degrees in a clockwise direction, for they flew according to their own, artificially shifted time. Matthews kept pigeons which previously had shown good orientation, for 6 days under conditions of irregularly alternating light and darkness, and fed them irregularly. As a result, their orientation was completely impaired, obviously because of a perturbed sense of time. An artificial daily rhythm with a "day" shifted by three hours was then imposed on the same pigeons. The anticipated shift in the initial direction of flight was found.

These experiments and a series of similar ones strongly attest to the correctness of the theory of navigation in which a necessary presupposition is the existence of a time sense in birds, and they thus attest directly to the existence of the sense of time.

BRIEF REVIEW OF THE PSYCHOLOGICAL LITERATURE ON THE SENSE OF TIME

The history of research on the sense of time is a long one. As the older works on psychology show (for example, Wundt 1903), the sense of time in man was a favorite subject of interest for a series of psychologists of the past century, an interest that had not diminished at the beginning of the 20th century. In experimental psychology a great quantity of facts and observations were collected, and it is difficult to find a factor whose influence on the sense of time has not been investigated. This extensive experimental material has, however, two major insufficiencies: On the one hand, there is the enormous collection of factors without deeper interrelationships which have very little relevance to the question of the mechanism or essence of the sense of time. On the other hand, it is a matter of an unsystematized collection, in which individual data or groups of data were obtained by the most varied methods, so that comparison of the various results and their systematic classification, particularly from a quantitative standpoint, is indeed hardly possible.

Methods of Investigation

The methods of investigating the sense of time are very diverse. That of temporal conditioning is considered the best but at the same time the most laborious; however, by choice of a suitable reflex as an indicator this method can be made entirely objective, even in man. In the present work, I myself used the method of temporal conditioning successfully. As follows from the preceding brief review of temporal conditioning, or more specifically, from a specialized reference collection (Dmitriev and Kochigina 1955), the major part of studies of this reflex was concerned with problems other than the sense of time. Trace-conditioned reflexes could also be used as a methodological basis for study of the sense of time.

The various psychological methods (Vierordt 1868, Wundt 1903, Benussi 1913, Pauli 1950, von Skramlik 1934a, b, 1935) can be divided into two large groups: the sensory method, in which various stimuli are presented to the experimental subject for judgment of their duration or sequence; and the motor method, in which the subject himself produces an interval according to the relevant instruction. In both cases the methods can be classified further according to the stimuli employed (auditory, visual, etc., also in combinations) or according to the method of production of the intervals by the subject (the most frequent being key-pressing, with or without sensory control). Another criterion for further classification of methods is the definition of the task to the subject: In a number of sensory methods, the subject either judges the applied stimulus absolutely or, more frequently, compares the duration of two stimuli. In other methods, the threshold interval is determined for two stimuli that can be differentiated from one another appropriately, and with respect to their sequence. In the motor method, the subject produces an interval either defined absolutely in advance, or by imitation of a presented stimulus (reproduction); at other times the intervals are produced arbitrarily or with the shortest interval (production). (By a rhythmic series of intervals here is meant a preferred, or maximal, tempo.) In addition, there exist numerous other special methods that lie outside the presently indicated scheme, as for example in the investigation of rhythms. Were we to consider further how varied the procedures can be, i.e., the processes selected in individual methods, we would see how great is the methodological variety for the investigation of the sense of time.

Principal Concepts

It is necessary to explain some principal concepts from the field of time sense which are constantly repeated in the psychological literature. The dissimilarities between subjective estimation of time and physical measures of time are generally indicated by unfortunate terms, such as for example that the subject "underestimates" or "overestimates" time. By underestimation is meant that state when the subject appraises a time interval as being shorter than it actually is, and conversely. Thus if, for example, a subject states that two stimuli of duration 1.5 and 2 sec in a particular circumstance are equal (as for duration), we conclude that the second stimulus is overestimated or that the first is underestimated. In the motor method, however, such an indication is unclear and the results must always be described in greater detail.

Much work has been devoted to investigating the so-called "indifferent," or "optimal," interval. It has been shown that there is a substantial difference in the perception of short and long time intervals (Meumann 1893, 1894, 1896, Benussi 1913, Wirth 1937). Short time intervals are overestimated or are reproduced as longer, and conversely for long time intervals. The duration of a stimulus that is intermediate between these extremes and that is judged comparatively the most accurately, is termed the indifferent interval. According to the circumstances, it varies from 0.7 to perhaps 10 sec (Vierordt 1868, Estel 1885, Mehner 1885, Glass 1888, Kahnt 1914, Woodrow 1930, Koehnlein 1934, and others). For very long intervals of time, the duration of which cannot be perceived as a single whole, another regularity holds. An upper limit is given for the "psychical present time"; according to various authors it is approximately 4 to 20 sec (Wundt 1903, Schulz 1927, and others). The principal value of this information consists in that it provides the possibility of a larger number of different mechanisms of the sense of time for shorter and for longer intervals of time, a possibility which is also well founded otherwise, as will be shown.

The pursuit of the lower limit of the psychical present time or moment (von Baer 1864, see von Skramlik 1934b) has been less fruitful. This limit is defined as the shortest duration of a stimulus that can still be perceived to have a duration, for in very brief stimuli the dimension of time has in some manner already become less evident. It was a question of some basis for a unit of the sense of time. Great significance in this connection has been attached to the study of the "critical fusion frequency," that is, the frequency of flickering light for which the impression of flicker disappears and the impression of homogenous gray appears (see for example the review by Simonson and Brožek 1952).

In studies of the sense of time the measurement of the so-called "time order error" ("Zeitfehler") has been frequently employed. According to Köhler (1923), who introduced the concept, the temporal order of stimuli (in addition to other effects, as for example the length of the pause between the stimuli) is decisive in the estimation of two successive perceptions. An error of the order is indicated as positive when in the process of estimation the first perception is accentuated or emphasized in comparison with the second one, i.e., when for identical stimuli the first seems more prominent (in an investigation of the sense of time, more specifically, longer); it is negative when the second of the two stimuli is preferred. The time order error can be expressed quantitatively as the ratio of the duration of the two stimuli which appear to be identically perceived. The work of numerous authors (Freeman and Sharp 1941, McClelland 1943, Tresselt 1944a, b) has shown, in addition to other points of interest, the particular regularity (Stott 1935, Philip 1940, 1947), that for stimuli of short duration the time order error is positive, for longer intervals negative, and the intermediate time durations agree with the time of the indifferent interval mentioned. Apparently the same phenomenon is being measured by both methods (Woodrow 1930).

The Significance of Different Factors

Psychological research has disclosed, in the first place, considerable individual differences in the ability to measure time. These differences concern not only the accuracy or variability of judgments but also a tendency to overestimation or underestimation, and consequently the indifferent interval and similar personal constants (Woodrow 1933, Koehnlein 1934). Thus, for the same individual there may be a significant variability in the course of a day and from day to day (Hawickhorst 1934). Some authors (Hawickhorst 1934, Harton 1939, von Skramlik 1934, 1935) also found a clear difference between sexes; they reported that females made more accurate judgments in acoustical tests, whereas males were more accurate in visual ones. To be sure, the differences were not particularly great.

The relationship of the sense of time to age of the individual studied is a significant one. Two different dependencies are apparent: In the first place, the sense of time develops in ontogenesis, so that in humans its inception can be observed only at approximately 4 years of age, the degree of perfection of adulthood being attained at 13 to 14 years (Binet and Simon 1916). In the main, it is correlated with the development of general intelligence. Secondly, the sensation of the constantly accelerating passage of time with increasing age is universally known; it is the psychological basis of the concept of biological time (together with the retardation of biological processes with age and the irreversible changes in all tissues in the course of aging)—see, for example, Carrel 1931, 1939. This phenomenon is sometimes stated as the law of Janet: For a constant duration of stimulus, the length of the subjective duration of the sensation is inversely proportional to the length of life already lived (Janet 1928). I believe that over and above the various psychological interpretations this dependence of the sense of time on age is one of the most important pieces of evidence for a metabolical basis for biological clocks (see under mechanisms of the sense of time).

From an analysis of the existence of a sense of time (p. 5), mention was made of different aids by which man is assisted in the judgment of time by experience. This point is perhaps more closely connected with the dependence of the estimation of time on the question whether or not the interval being judged is "filled" or "empty," that is, whether it is determined by two brief stimuli or by a single stimulus that persists throughout the interval being investigated. It has been consistently found by various authors (for example, Glass 1888, Meumann 1893, 1894, 1896) that a filled interval is rather overestimated in comparison with an empty one (at least for intervals of up to 2 sec). Further, the intensity of the stimulus influences the estimation of time (Meumann

1893, 1894, 1896, Woodrow 1909, Quasebarth 1924, Hormia 1956), such that stronger stimuli lead rather to overestimation or a positive time error.

Many studies have been devoted to the question of the effect on the sense of time of those psychical factors that are ordinarily included in the concept of attention, or "attentive set" (Ejner 1889, Meumann 1893, 1894, 1896, Schumann 1893, Wundt 1903, von Kries 1913, Hülser 1924, Schulz 1927, Jaensch and Kretz 1932). Quantitative data are of little value, for there is a good deal of controversy involved here. For example, it was said that focusing of attention leads to an underestimation of time (Wundt 1903), but the same was said for lack of focusing of attention (Hülser 1924); thus two attentive sets were distinguished, active and expectant, with contrary effects on the estimation of time, etc. What is important, however, is that the subject can, by deliberation and intentional intervention, significantly influence his estimation of time, distorting his own experimental results by which the sense of time was being investigated. The significance of the method of temporal conditioning for the present research is therefore emphasized, for only by this method can such pitfalls be confidently avoided.

Other aspects of the subjective relationship to the estimation of time have a greater influence on its accuracy. Harton (1938, 1939, 1942) studied the effect of the ease or difficulty of an assigned task on this sense, the effect of uniformity or diversity of the task, and the influence of success or failure or fear of failure. Here it is already a question of emotional influences or affective coloring of experiences, which greatly influence the correctness of the estimation of time (see, for example, Schulz 1927); intensive experiences to the point of so-called psychical shocks can result in a transitory complete extinction of the perception of time (Ferrari 1909). According to Gulliksen (1927), pleasant psychical states result in underestimation, unpleasant ones in overestimation of time. Upon retrospection, however, it is the other way around. Enormous differences between the actual flow of time and its subjective perception can be brought about by appropriate suggestion under hypnosis (Cooper and Erickson 1954). From similar descriptions in the literature it would be possible to continue on at some length; we shall, however, limit ourselves to the reports mentioned, adding that there is also the effect of fatigue, although on this point there is no unanimity among the results obtained (Ejner 1889, Schulz 1927).

On methodological grounds, it is further necessary to direct attention to a body of information concerning the significance of the modality of the stimuli and similar factors on the estimation of time. In the main, the best results have been obtained for auditory stimuli (von Skramlik 1934a, b, Homack 1935, Woodrow 1951). Further, when the intervals to be estimated are specified by brief stimuli presented from different sides, or stimuli of different modalities, the estimation of time deteriorates in comparison with a uniform stimulus specification of these intervals (Grassmück 1934, von Kries 1923). The role of proprioception in the estimation of time has also been emphasized (Münsterberg 1889, Guilford 1929). Indeed the observation of Münsterberg, that perception of time is intrinsically the perception of change of muscular tension is but poorly explained.

In a series of identical tests carried out repetitively, the ability of estimating time at a constant level is frequently not maintained but changes in the course of the measurement: According to Ejner (1889) and Schulz (1927) there is usually an initial underestimation which alternates transiently with an overestimation, following which an underestimation again sets in. Hormia (1956), on the contrary, stated that there was a general tendency toward overestimation at the beginning of a series of tests. Also, test stimuli of varying duration, applied in a temporal sequence, mutually influenced their subjective time value significantly (Glass 1888, Israeli 1932, Philip 1947). Thus, for example, a stimulus seems longest when it is preceded by a shorter and followed by a longer stimulus.

The Basis of the Sense of Time

Concerning the mechanisms used by organisms in the measurement of time, i.e., the basis of the sense of time, a thought already apparent to the early psychologists (Münsterberg 1889, Ejner 1889, Wundt 1903) readily occurs, that it could be some kind of rhythmic process, or rather a rhythmic process occurring in the body such as, for example, respiration, the heart beat, etc., which could subserve the sense of time analogously to the pendulum in a clock. Therefore, one can also speak of a biological pendulum. Actually, people under the pressures of everyday life estimate time without technical aids, frequently making use of rhythmic counting, tapping the hand, etc., in other words creating artificially some kind of an improvised reference rhythm. As von Skramlik (1934a) noted, people behave similarly during laboratory experiments in the field of the sense of time, for example, counting their respirations, unless the latter has been forbidden in the instructions beforehand. As was mentioned in the survey on temporally conditioned reflexes, such an interpretation of the mechanism of such reflexes (Pavlov) has thus far been the best one. A certain grotesque culmination of such views is the concept of Gooddy (1958), according to which the entire human organism is nothing other than a clock, since it consists of a series of autonomous rhythms and in this way conforms to an

Einsteinian definition of a clock.[1] Nevertheless, it seems profitable to survey other material in the literature on this point.

Goudriaan (1922) investigated in man the reproduction of a rhythm (estimation of a time duration of 20 sec) as well as a preferred tempo, and correlated the results with the subjects' resting heart and respiratory rates. In addition, he examined the influence of acceleration of respiration and heart beat on the results by having the subjects perform deep knee bends. The only affirmative result was the finding of a positive correlation between resting heart rate and the preferred tempo of the subjects, such that subjects who preferred a faster tempo overestimated time, whereas for slower tempos there was an underestimation. The results of other authors (Schaefer and Gilliland 1938) in this area were on the whole negative.

More conclusive evidence for the existence of a "biological pendulum," although less specific (as concerns a specific kind of biological rhythm that could serve the purpose), are reports of a positive correlation between the sense of time and the rate of metabolism, for in the final analysis there can be no doubt that all biological rhythms arise as an interaction of individual chemical reactions of metabolism. Relevant to this point is the previously mentioned slowing of biological time (p. 19), which

---

[1] "What is a clock?

"The primitive subjective feeling of time flow enables us to order our impressions, to judge that one event takes place earlier, another later. But to show that the time interval between two events is 10 seconds, a clock is needed. By the use of a clock the time concept becomes objective. Any physical phenomenon may be used as a clock, provided it can be exactly repeated as many times as desired. Taking the interval between the beginning and the end of such an event as one unit of time, arbitrary time-intervals may be measured by repetition of this physical process. All clocks, from the simple hourglass to the most refined instruments, are based on this idea." (Einstein and Infeld 1938)

parallels the retardation of organic processes and the aging of tissues with increasing years. Experimentally elevated temperature in man (by means of general diathermy) speeds up a freely produced tempo, entirely according to expectation. The Q_{10} of this effect was 2.75 to 2.85 (François 1928). Hoagland (1935) found a similar dependence of the sense of time on body temperature; for the acceleration of electroencephalographic rhythms in man with body temperature he even found a mathematical expression (Hoagland 1936). Surprisingly, Folk et al. (1958) found an independence of the "biological clock" and temperature in hibernation. Grabensberger (1933) altered the biological rhythms of ants by means of temperature, consonantly with the effect on their metabolism. Metabolic poisons that slow metabolism (salicylic acid, phosphorus, arsenic) slowed living rhythms of ants (Grabensberger 1934b). Iodothyroglobulin, which accelerates metabolism, accelerates living rhythms of bees and wasps (Grabensberger 1934a). Even in man the effects of pharmacological agents on the sense of time partly support the hypothesis of a metabolic pendulum in the organism. Thus Sterzinger (1935, 1938) reported an underestimation of time with quinine, an overestimation with thyroxine, caffeine, and theobromine, and he stated that the preferred tempo correlated with metabolism. On the other hand, the findings with alcohol are very different and contradictory; here it may be a matter of a nonspecific disturbance of the sense of time. As is well known, mescaline and hashish (Vondráček 1935) especially have a prominent effect on the perception of time, where it is already a question of hallucinations.

From human pathology, surprisingly, no results have been reported from the investigation of time sense in thyrotoxicoses nor in myxedema (Gardner 1935). Injuries to the brain and neurological experience generally attest, it is said, to the importance of the thalamus and hypothalamus for time sense (Straus 1928, Ehrenwald 1923, 1931). According to Vondráček (1949) the perception of time is localized in the mammillary bodies and great significance is attributed to the cortex, especially of the temporal lobes. Spiegel et al. (1955) found transient disorders of temporal orientation and estimation of time in 23 out of 30 patients with operations on the thalamus (dorsomedial thalamotomy). They concluded that connections between the dorsomedial nuclei of the thalamus and the frontal lobes are probably important for the sense of time. Numerous disorders of the sense of time have been described in various psychoses (bizarre disturbances, for example, are well known in Korsakov's psychosis: Perception of the present is intact but the past and the future lose their significance and the sense of the continuity of time is lost). Altogether, however, there is very little in this area of a specific nature which could cast light on the mechanism of the sense of time.

It is of course necessary to point out that neither the substantiation of the existence nor the identification of individual pacemakers for time, both of which are of themselves far from being accomplished, would signify a complete resolution of the problem of the sense of time, for there would remain to be explained how cycles of the respective rhythms are counted and converted into different time segments in the organism.

THE WORKING HYPOTHESIS AND PLAN OF ITS EXPERIMENTAL VERIFICATION

The organization of the sense of time is conceived of in such a way that as its basic mechanism there is some kind of automatic rhythm which occurs continuously in the body and with which the organism can compare the duration of stimuli or its own movements. This reference rhythm must be autonomous, that is, not directly derived from rhythms in the environment. Further, its variability must be comparatively small—for example, the heart beat or respiratory rhythm would not be suitable because of their great dependence on a large number of factors, causing their rates to vary over very wide ranges. Finally, the autonomous rhythm must have a suitable frequency. This latter requirement leads to the conclusion that in the body no single rhythm can serve as the biological pendulum of a clock, but rather that there must be a number of such rhythms, and that in a particular instance the one whose frequency would be commensurable with the time being measured would always, or predominantly, be invoked. Thus, it cannot very well be imagined that time intervals of the range of days or weeks would be measured by a rhythm of a frequency of, for example, several cycles per second, for a day lasts 86,400 seconds. Indeed, in living organisms we encounter the principle of multiple assertions of different functions everywhere, and several such temporal rhythms in the body could further enable their mutual control. Of course, in a specific, concrete case, only one single rhythm would as a rule be used as a biological pendulum.

On this point, some general remarks concerning biological rhythms should be added.

Biological Rhythms

The existence of very different rhythms in the human and animal organism is well known. The range of frequencies of these rhythms is enormous. The electrical activity of the brain manifests rhythms of frequencies up to some tens of cycles per second. Some metabolic or endocrine functions display a distinct annual cycle. Many biological rhythms show a direct dependence on natural physical rhythms (especially daily and seasonal rhythms), on cyclical changes of light, temperature, etc. These rhythms can no doubt with long development become so fixed in the organism that they can be altered only with great difficulty by changing the external rhythm. After all, however, they truly represent an adaptation (Webb and Brown 1959) of the organism to these external cycles of changing conditions, so that their origin is primarily in the rhythmical changes of the environment.

Other biological rhythms, on the other hand, represent automatisms of the organism or some part of it, and hence it would be futile to search for their dependence on rhythms of the environment. For example, the rhythm of the heart is entirely autonomous; it beats with its own regular rhythm for a long time after removal from the body, under suitable conditions. I have maintained excised hearts in activity in vitro for several days (Holubář 1956), in experiments designed for quite another purpose. Just such an autonomous rhythm could serve as a biological clock which could enable the organism to measure time and in those cases when there is complete isolation (in the main, under artificial and experimental conditions) from those influences of the environment which could otherwise assist in the measurement of time. (The automaticity of such biological rhythms must be properly understood. By no means does one speak against the unity of the organism with its environment, or even against determinism. There is nothing at all mystical about autonomous rhythms of the organism.)

We shall analyze a simple physical rhythmic process, the falling away of drops of water from an opening in a vessel. A drop grows by the flow of water until a definite size is reached, whereupon it breaks loose, and the process repeats itself. The growing drop clings to the opening by surface tension, which increases with enlargement of the drop proportionally to the square of its radius (if for simplicity the drop is considered a sphere). Since, however, the weight of the drop increases in proportion with the cube of the radius, the weight periodically exceeds the surface tension, and at this point the drop separates.

This physical rhythm is in no wise a model for biological rhythms but is only illustrative to show that rhythms can arise in nature from the interaction of two simple continuous processes, with the linear growth of two forces— in this case gravitational and surface tension. Thus, in an instance as clear as is the formation of drops, the whole phenomenon can be described quite readily mathematically and the manner in which the frequency of the rhythm depends on the size of the opening, for example, can be expressed. Also it is at first sight evident that the rhythm with which the phenomenon occurs is not given by any rhythm in the environment. Thus, it would be surprising enough if no intrinsic rhythms, i.e. rhythms quite independent of the environment, would occur in the organism, where mutually closely related chemical reactions and physical processes are always evident.

The rhythm of the heart beat itself is a vivid example of such an autonomous rhythm of metabolic origin (accelerating with increased temperature), influenced by the nervous system and the chemical constitution of the internal environment in response to the most diverse factors (physical work, emotion, etc.), where in substance it is a matter of a rhythmic oscillation of the excitability of the sinus node, which is in no wise linked to a rhythm of the external environment.

Brain Rhythms

A further part of my working hypothesis is the supposition that for times of the order of seconds to minutes (the intervals most frequently studied under laboratory conditions) in higher animals, specifically in man, one of the rhythmic activities of the brain could be a time-measuring rhythm, so that a pacemaker for the sense of time would be located in the brain, or in some part of it. Brain rhythms, well known from studies of the electrical activity of the brain (electroencephalography—see, for example, Hill and Parr 1950) are truly autonomous; that is, they are not based on any rhythm of the environment and they are of a comparatively constant frequency. For example, the frequency of the alpha rhythm is a sort of personal constant, not being under the control of external influences, unchanging with age in adult subjects, and its frequency spectrum behaves accordingly. The requirements mentioned at the beginning of this chapter would thus be met. Moreover, this conception conforms to the dominant position of the central nervous system, whereas if the localization of the supposed rhythm were in some other part of the body, sensitive connections with the central nervous system would still be essential.

An important feature of this conjecture is that it offers a way for testing it. A peculiarity of brain rhythms is that, in spite of their practical immunity to external influences, a means exists of altering them at will quite substantially and quantitatively to a significant degree: Optical stimulation of the eyes with interrupted light of a suitable frequency results in a synchronous rhythm of electrical activity in the brain which rapidly spreads over practically the entire brain (Adrian 1947). This phenomenon has been extensively studied, initially in the rabbit (Bartley and Bishop 1933, Bartley 1936, 1940), then in the monkey (Halstead et al. 1942a, b), and especially in man (Adrian and Matthews 1934, Durup and Fessard 1935, Loomis et al. 1936, Jasper 1937, Toman 1941, Walker et al. 1943, 1944, Walter and Walter 1949; see also Holubář and Kohlík 1950). It can then perhaps be expected that flickering light would also alter the rhythm of a center that is decisive for the sense of time and that as a consequence, under favorable circumstances, the sense of time would either be impaired, or even specifically influenced. An affirmative result would support the working hypothesis; it would even attest to the point that for the appropriate time intervals there exists a unique pacemaker in the brain. On the other hand, a negative result would not signify anything; the working hypothesis would not be contradicted, for the supposition could readily be advanced that the critical part of the brain remained unaffected by the imposed rhythm. (To be sure, it could be a matter of a microscopically small part.)

The effect of flicker on electroencephalographic rhythms is specific also in the sense that the flicker does not affect other functions of the organism. For example, at the rates necessary for our experiments, it certainly does not influence metabolism, in contradistinction to the effect of pharmacological agents, changes of temperature, etc., which as a rule have a significant effect on a whole series of functions of the organism; their effect on the sense of time, which has been unquestionably demonstrated, must therefore be interpreted for the most part as significant only generally, rather than in any specific way.

Plan of the Experiments

For carrying out the experimental verification of the working hypothesis, it was first necessary to select an appropriate test of the sense of time. In the preliminary experiments, the reproducing of a rhythm, and after some experience with modification of it, the free production of a rhythm, were selected as simple methods that were not very demanding and yet suitable for orientation. When the preliminary experiments gave promising results, as objective as possible a method was sought. After examining the method of trace-conditioned reflexes, which did not prove suitable, that of temporally conditioned reflexes was considered; this was found to be very challenging, to be sure, but also very thorough, objective, and conclusive. As an indicator, the galvanic skin reflex was selected (Tarchanoff 1890, Veraguth 1909) and in all of the main experiments, also the EEG (several leads simultaneously). In addition, respiration and the EKG were recorded in a part of the experiments. The galvanic skin reflex and EEG proved to be especially suitable as responses that could not be intentionally affected by the subject, and were thus entirely objective indicators. Both the galvanic skin reflex and changes in the EEG accompanying its establishment were recorded in a series of preliminary experiments, and were used to examine a number of subsidiary questions in the course of the main series of experiments. The presentation of flicker was carried out in the preliminary experiments by means of an imperfect arrangement; in the main series, an adequate photostimulator was employed. The flicker was always applied only for a short time of 1 to 2 minutes, in part so as not to evoke undesirable secondary effects, such as fatigue, etc., and in part because its effect on the EEG is always greatest at the beginning of an application, diminishing in the course of time. In the main experiments, a constant frequency of flicker was always used in a given application, the frequency being changed only in another application, so that the effect of the particular frequency of flicker under study would be as clear as possible. As a control for any possible nonspecific effect of light, a steady light was also used. In order to meet the objection that perhaps flicker influences the sense of time primarily by means of an effect on the rhythm of breathing or the heart beat, the question of such possible effects was also examined in rabbits. Otherwise the entire investigation was carried out on human subjects. In the question of the sense of time, man is probably of the most central interest; results gained from animals, although of great importance and interest, constitute another and to a significant degree independent chapter.

METHODS

Experimental Subjects

Excepting some complementary experiments on rabbits, measurements were carried out entirely on clinically healthy human subjects. Altogether 29 subjects were used, among whom 4 were studied in the preliminary experiments with reproduction of intervals and with freely chosen tempos, another 10 in preliminary experiments with recordings of changes of skin resistance, and a further 15 subjects in the main experiments with recordings of skin potentials and EEGs. Subjects were 17 to 35 years of age, the majority around 20 years; 7 females and 22 males were used. They were principally students, 12 being medical students, 13 students from other departments (industrial, advanced economics, etc.), and 4 of other occupations.

The subjects were not advised of the purpose of the experiments, and the instructions to them were as brief as possible. The preliminary and preparatory experimental series were performed in a darkened room where the subjects sat in a comfortable armchair; the experimentor was as far removed from them as possible, additional separation being obtained by a screen. The principal experimental series was carried out in such a way that the subjects sat in an armchair in a soundproof room, in dim light (but not in complete darkness), and so that the experimentor could be addressed at any time by means of an intercom; the experimentor could speak to the subject only by pressing a microphone switch. Subjects were inconspicuously requested to remove their watches by requiring that the left wrist be used as the ground electrode. Care was exercised so that no session ever lasted more than 2 hours (including the entire preparatory and final cleaning up), and so that sessions did not follow one another within less than 48 hours. As a preliminary to an experiment, subjects were accustomed by at least one preparatory session without the presentation of stimuli.

Apparatus and Experimental Procedures

Nociceptive stimulation consisted of paired electrical shocks of DC current from a "B" battery. The voltage employed was 80 to 120 volts, the duration of the shocks was very short, 0.5 to 1.0 msec, the current being switched on by means of a special key with a short time of contact (Holubář 1949); the two shocks were separated by an interval of 0.2 to 0.3 sec. In this way it was possible to achieve an appreciable stimulus, to the point of being unpleasant but not definitely painful. The stimulating electrodes were silver discs with a diameter of 1 cm placed on the palm and on the dorsum of the hand with the aid of EKG jelly. Besides the fact that this kind of stimulus was the most suitable for our purpose, the arrangement had two further important advantages: First, there was no risk of accidental injury to the subject from the electrical current, for the entire simple circuit could be inspected at a glance and there was no connection with the electrical network system. Secondly, the arrangement permitted the delivery of stimuli simultaneously with the EEG recording without interfering with the EEG record. The stimulus artifacts appearing in the record were a useful indication of the instant of stimulation; when they were too large, they could be diminished by applying a square electrode (a strip of lead sheet 3 × 8 cm covered with gauze and soaked in physiological salt solution) to the wrist of the extremity being stimulated and connecting the electrode to ground.

The flicker was obtained in the preliminary experiments by means of a small projector (5 × 5 cm), in front of which rotated a disc with sectors removed, which was masked by a black screen with an opening for the light. The light source was placed 1 m from the eyes of the subject. The frequency of flicker was changed by means of a regulating transformer supplying current to the electric motor which drove the sectored disc. In the experiments of the main series, a far more sophisticated photostimulator (a Kaiser stroboscope) was used, in which the light source was a discharge tube, and the flicker was achieved by means of an electronic circuit. The reflector of the stroboscope was likewise 1 m distant from the eyes of the subject, a light intensity of 0.2 joule being used. For a constant light (as a control) a reflector was used which resembled the reflector of the stroboscope, but instead was provided with a 100-W incandescent lamp.

The recording of the galvanic skin reflex was carried out in two ways. In the preliminary experiment, the change of skin resistance was obtained with the aid of a bridge circuit of my own design (Fig. 1). The readings were taken visually from the screen of the cathode-ray oscilloscope. The bridge was always initially balanced to give the minimum width of the line on the screen, by means of adjusting the potentiometer and selecting the appropriate capacity of the condensor (0.1 to 0.2 microfarad). The reflex was expressed as a widening of the line, whose width sometimes did not return to the original value. It was thus necessary in the course of experiments

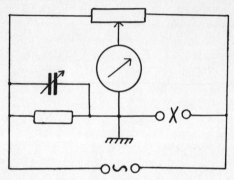

Fig. 1. Impedance bridge for observing changes of skin resistance. The recording electrodes applied to the skin are connected at X. The bridge is fed by an AC current of approximately 2 volts, balance is indicated on a cathode-ray oscilloscope. The bridge is balanced by means of the potentiometer (10 to 500 kΩ) and a stepped capacitor (0 to 0.25 μF, in steps of 0.01 μF). The fixed resistor is 10 kΩ.

to make readjustments gradually (i.e. to rebalance the bridge) both by means of the potentiometer and by changing the capacity, usually by increasing them.

In the experiments of the main series, the same arrangement could not be used because the auxiliary measuring current appeared in the simultaneously recorded EEG, even when the bridge was fed from a special oscillator of 200 Hz frequency. Therefore the change of skin potential was recorded without the auxiliary current in the latter series, by connecting the electrodes directly to the input of one (the seventh) channel of the electroencephalograph. Inasmuch as the maximal time constant of the amplifier was 1.0 sec, the course of the change of skin potential was of course distorted, but this was immaterial for our purposes, since a reliable indication of the existence and the time of onset of the potential change was sufficient, and our apparatus showed this adequately.

The electrodes for recording the galvanic skin reflexes (i.e., recording changes of resistance and of potential) were the same in both cases. They were two silver discs of 3.5 cm diameter, one convex and the other concave, which were applied with the aid of EKG jelly to the palm and to the dorsum of the hand, respectively, so as to have close contact with the surface. The surface of the silver electrodes was cleaned (with fine emery paper) to a metallic luster before each experiment; the skin was preferably neither prepared nor washed before an experiment, for otherwise the period required for stabilization of resistence or potential was longer.

The EEG was recorded directly onto a Kaiser writing instrument. Of the 8 channels, one was required for the skin potential and on occasion another one for recording respiration or the EKG, so that either 6 or 7 EEG channels were recorded. Symmetrical bipolar recordings were made from 3 leads disposed on each side of the head. The electrodes were located longitudinally and for the most part there were 3 on each side, of which the third linkage was lengthwise fronto-occipitally, so that the imposed rhythms could be expected to be maximally distinct. On later occasions, this point of view was abandoned and 4 electrodes were placed on each side. The linkages are shown schematically in the figures. The electrodes used were always silver discs of 10 mm diameter, covered with gauze soaked in physiological salt solution, EKG jelly being applied to the skin under the electrode. The electrodes were held in place by means of an improvised cap consisting of narrow rubber bands, which were connected at the crossing points by means of special buttons. The electrodes were located such that their contacts (little tubes fastened perpendicularly to the center of the discs) were spaced in holes in the rubber bands. This method proved satisfactory not only from the standpoint of stability of the fixation, but also because it did not inconvenience the subjects, even during two-hour sessions.[1]

The EKG was recorded in some of the experiments. Since only the pulse rate was of interest, the ear and the opposite hand were usually used for the placements of the electrodes. Respiration was also recorded in some experiments. This was accomplished by means of a pneumograph connected to one channel of the EEG (Holubář 1959b). The recording pneumograph was wrapped around outside of the clothing and in no way inconvenienced the subject.

---

[1] The EEG electrodes and their method of application were based on the suggestion of Dr. Drechsler of the Neurological Clinic in Prague.

RESULTS AND THEIR ANALYSIS

Preliminary Series

In the preliminary series of experiments, 4 subjects were used. They sat in a darkened room, in a comfortable chair, with their hand on an electric push button. Each pressing of the button was signaled by means of an indicator light and was recorded. Two kinds of experiments were carried out: reproduction and production of a rhythm at a freely chosen tempo. The reproduction was tested in such a way that the experimentor initially demonstrated to the subjects 5 equal intervals, of 1, 3, 5, 10, or 15 sec (by means of the lighting of a small incandescent bulb), after which the subject immediately began to reproduce the interval by pressing the button until stopped by the command, "sufficient" (i.e., about 10 times). After a short pause, the subject repeated the procedure in the same way. Initially, all experiments were carried out only in darkness; later flicker was used in some experiments (but only in that part of the experiment in which the subject himself reproduced the intervals).

Production of an interval at a freely chosen tempo was examined in such a way that the subject pressed a button in an arbitrary rhythm for the whole duration of the experiment, having been given only the instruction that care should be taken to maintain constancy of intervals. At the same time, flicker was applied intermittently for periods of approximately 15 to 30 sec, alternating with periods of darkness, without an intervening pause.

Although the results in the two kinds of experiments were in agreement, the clearest was in the case of the freely chosen tempo, in two respects: the interval varied only slightly in darkness, for example between 5 and 7 sec, and the effect on the interval of flicker was more striking. The most effective flicker was that of a rapidly changing frequency (for example, with an ascending course over half a minute from 4 to 14 per sec, or descend-

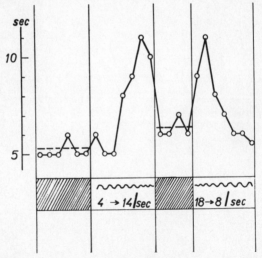

Fig. 2. Effect of flicker on spontaneously produced tempos. The shaded portion denotes trials in darkness, the waves under the graph denote trials during flicker, the frequency of which in the first instance increased, and in the second instance decreased. The intervals, which varied between 5 and 7 sec, were prolonged to as much as 11 sec.

ing in the same interval from 18 to 8 per sec). Figure 2 shows a typical example of the effect of a flicker of changing frequency on a freely produced interval. The intervals were prolonged to as much as 12 sec as compared with a control of 5 to 7 sec. The results in all experiments of the preliminary series were the same or nearly so. It can then for the moment be concluded that flicker clearly affects the length of the interval which the subject freely produces or reproduces.

The effect of flicker in these preliminary experiments was always such that the interval was prolonged. Mention will be made subsequently of experiments in which, by contrast, the intervals of temporally conditioned reflexes were shortened by flicker. These two results are not actually contradictory, for the difference can be explained such that the effect on the specific function of the sense of time, i.e., on the measurement of time, was the same in both cases but dissimilarly expressed. In the case of temporal conditioning (see p. 62) the subject was quite unaware of the fact that he was producing certain intervals and therefore the speeding up of his "biological clock" was manifested directly as a shortening of the intervals. In the case of the preliminary experiments, a subject was conscious of the speeding up of the biological clock and in attempting to maintain the intervals constant, compensated the error of his "pendulum" in such a way that the result was a prolongation of the intervals.

The principal result of these preliminary experiments was to reassure us in pursuing our working hypothesis further and to justify an attempt to verify it reliably. In other respects, the significance of the preliminary experiments should not be overestimated. There existed a certain element of uncertainty in the fact that the subjects had the possibility of intentional interference. There is perhaps a very simple interpretation of the effect of the flicker on the intervals, i.e., by diversion of attention which, it must be admitted, would say very little about a genuine mechanism for the phenomenon and would undermine the significance of the experimental results for our working hypothesis. The photostimulator used for the preliminary experiments was imperfect (it did not permit a constant frequency of flicker to be maintained); this interfered with the possibility of a more accurate quantitative study of the phenomenon. Nonetheless, because of the other deficiencies of the method in the preliminary experiments, they were not repeated with an adequate photostimulator when the latter became available, since by then the more satisfactory method of temporally conditioned reflexes had been taken up.

Galvanic Skin Reflex (with the Aid of an Auxiliary Current) and Its Conditioning

The change of electrical resistance in one hand was observed in 10 subjects (5 males and 5 females) during 56 sessions altogether. As the unconditioned stimulus, either paired electric shocks to the other hand (see section on methodology) were used, or loud sounds (horn, bell). The sound of a buzzer lasting 3 sec was used as the conditioned stimulus. The unconditioned stimulus was applied at the termination of the conditioned stimulus. For a reliable evaluation of the conditioned reflexes, it was found to be advantageous to use for the conditioned stimulus a sound of the buzzer sufficiently weak so that the latter did not of itself result in any change of skin resistance. Stronger sounds alone resulted in a response in the form of a variation of skin resistance, and it was always necessary first to extinguish this orienting reflex by several presentations of the conditioned stimulus alone; otherwise the objection could always be raised of the question of a reestablishment of an orienting reflex, as against the elaboration of a conditioned reflex.

Experience acquired in this series of experiments did not in general entail any new findings, to be sure, but for us they were of great significance as preparation for the main experimental series. First of all, it was shown that there is great variability in the occurrence and intensity of the unconditioned galvanic skin reflex in different subjects, and even in the same subject under different circumstances. Comparatively rare were the extreme cases of complete inability to carry out the experiment, i.e., on the one hand complete absence of any kind of change of skin resistance to even the strongest stimuli, and on the other hand spontaneous variation of skin resistance for the entire time of a session, without apparent external cause. The latter case was for the most part associated with increased perspiration of the hands. Absence of a galvanic skin reflex (i.e., of a change of resistance) was noted especially in the period of a few days before or after influenza. This interesting observation was confirmed anew in later experiments, by a change of skin potential (see p. 60). This phenomenon is perhaps primarily vasomotor in origin, for frequently (but not always) a skin galvanic reflex returned upon immersion of the hand in hot water for 5 to 10 minutes, followed by drying.

The galvanic skin reflex was found to be a very advantageous method, on the whole, for the study of conditioned reflexes in man. It was readily demonstrated, without difficulty for the subjects, who were not in any way inconvenienced by the recording of the necessary experimental data. Subjects could not consciously interfere with the results, nor did they know what or when something was being produced in them. Finally, the galvanic skin reflex was readily conditioned. In our arrangement it was not infrequent that a conditioned reflex appeared even after three reinforcements and shortly was so well established as to be difficult to extinguish. This point concerns of course the usual process of conditioning.

In this series of experiments, temporally conditioned reflexes were produced in such a way that only the unconditioned stimulus was presented and repeated many times at regular intervals (for some subjects, 30 sec, for others, 60 sec). The galvanic skin reflex then appeared several times at the same interval, without any kind of external stimulus, following the termination of the series of stimuli. For elaboration of a consistent temporally conditioned reflex, many reinforcements (100, up to a few hundred) were necessary, usually during several sessions. To be sure, indications of a developing temporally conditioned reflex were unmistakable much earlier, after several tens of reinforcements. The results here were similar to those described for the main experimental series. The greatest virtue of temporally conditioned reflexes for our purpose was the great accuracy and the small dispersion of the values of the intervals, especially for reflexes that were firmly established. This attribute was then employed in the main series (see p. 69).

Trace-Conditioned Galvanic Skin Reflex

These were studied in three subjects. Changes in the electrical resistance in the hand were examined, paired electric shocks to the other hand being used as the unconditional stimulus, as described under the heading Methods. The conditioned stimulus was a 2-sec sound of a buzzer of sufficiently low intensity so as not to result in a change of skin resistance as a component of an orienting reflex. The unconditioned stimulus always followed the beginning of the conditioned stimulus by 10 sec, so that between the stimuli there was a silent interval of 8 sec. The conditioned reflex was then elaborated immediately as a trace reflex, no gradual separation being used.

The establishment of a trace-conditioned reflex was found to be comparatively easy; thus, in one case the conditioned reflex appeared three times in succession even after 15 reinforcements. At another time it appeared 5 times after 35 reinforcements. As a rule, however, several sessions and about 100 reinforcements were necessary. It was found that for a consistent trace reflex the interval was not sufficiently constant, varying over a wide range between 5 and 20 sec, and there was no perceptible tendency toward a gradually increasing accuracy. It was apparent that such variation of the interval could not be used successfully for the study of its eventual change under the influence of flicker, and therefore trace-conditioned reflexes were abandoned as a method of approach to our problem.

My results with trace-conditioned reflexes are in general in agreement with the findings of Leonow (1926).

Main Experimental Series

Altogether 233 sessions were carried out on 15 subjects (1 female, 14 males) (Table 1). The results with the galvanic skin reflex, which was recorded without an auxiliary current (i.e., as changes of skin potential) were in complete agreement with the results of the preliminary experiments (in which changes of resistance were recorded), so that it is not necessary to repeat them anew. In addition to the effect on the intervals of the temporally conditioned reflexes, which was the most important aim of the entire series of experiments and concerning which there will be a discussion at the end of this chapter, some observations and findings were made which were in part closely related to the main theme, and which will be interpreted below. These concern the changes in the EEG that accompany the unconditioned and conditioned galvanic skin reflex, instances of dissociation of the galvanic skin reflex and such EEG changes (in particular with regard to change of skin sensitivity), some information on temporally conditioned reflexes (i.e., concerning

Table 1. Summary for Subjects in the Main Series

Subject	No. of Sessions	Temporally Conditioned Reflexes	K Waves[1]	Alpha Frequency[2]	Effect of Flicker[3]	Graph (Fig. No.)	EEG (Fig. No.)
PR	45	+	+	10.4	p	23, 24	4, 5, 6, 10, 11, 15, 16, 17, 18, 21, 22
VV	23	+	+	10.0	p	25, 26	8, 19
KL	17	+	+	10.5	p	27, 28	20
UV	10	+	+	10.75	p	29, 30	
CH	22	+	+		n		3
PV	18	+	+		n		
IV	12	+	+		n		7, 9
JG	22	+	+		0		12
JK	11	+	+		0		14
CR	8	+	+		0		
KP	7	+	0		0		
BR	15	+	0		0		13
VS	11	+	0		0		
DF	5	+	0		0		
SK	7	0	0		0		
15	233	14 93% Total	11 73%				

[1] Occurrence in general, i.e., as an unconditioned or conditioned reflex
[2] Average frequency (in c/sec) of the alpha rhythm computed from 10 sessions
[3] p–effect of flicker conclusive; n–effect inconclusive because of small number of successful experiments; 0–either only suppression of the conditioned reflex, or not examined

the question of their elaboration, inhibition in the course of an interval, and the mechanism of their origin), and finally, the finding of a conditioned reflex in the occurrence of K waves.

EEG Changes Accompanying the Galvanic Skin Reflex
(Holubář 1958a, 1959a)

1. It was frequently possible to note a slow fluctuation of potential in the EEG record, which was very similar in form to the galvanic skin reflex recorded from the hand and had a time course approximately similar to the latter, but was of a somewhat shorter latency, i.e., the galvanic skin reflex on the skin of the head. It was most prominent in the frontal leads, but was frequently apparent over almost the entire head including the occipital leads. Its nature, to be sure, was not related to the EEG. It is well known to electroencephalographers as a slow oscillation that interferes with the EEG recording in some subjects, and is eliminated from the EEG recording by decreasing the time constant of the amplifiers. The phenomenon is clearly apparent in Figs. 3, 4, and 5.

2. The K wave, or K complex, was described by Loomis et al. (1935a, b, 1937, 1938); further information on it can be found in the papers of additional collaborators (H. Davis et al. 1937, 1938, 1939a, b, c; P. A. Davis 1939a, b) and more recently in the paper by Roth et al. (1956). The K wave is a comparatively slow monophasic, biphasic, or multiphasic wave of high voltage (a single one lasting 0.2 to 0.3 sec or more). In the so-called K complex, it is followed by a spindle of waves of frequency of around 14 per sec. The K wave or complex is found in man in response to auditory or any other kind of sufficiently strong stimulus, with a latency of 40 to 170 msec, bilaterally and generalized over the entire head, typically in sleep, but also in the waking state.

48 RESULTS AND ANALYSIS

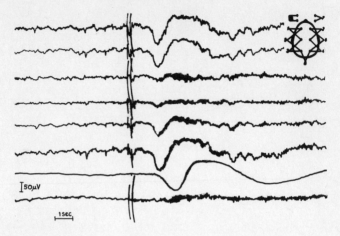

Fig. 3. Prominent unconditioned galvanic skin reflex on the scalp and appearance of 14/sec spindles in the EEG. The EEG linkages are indicated at the upper right. The skin potential from the hand is recorded in channel 7. Stimulation artifacts in all leads indicate the instant of presentation of the stimuli. The same notation is employed in the figures following.

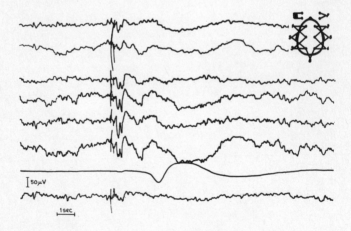

Fig. 4. Unconditioned reflex. The skin response is also present on the scalp, and a K wave is present in all EEG leads.

In my experiments, K waves (less commonly K complexes, see Figs. 9 and 14), were very often found in responses to nociceptive stimuli and were consequently

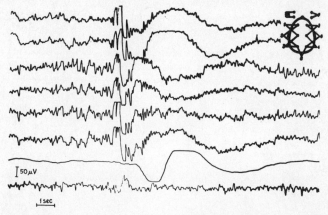

Fig. 5. Temporally conditioned reflex (third in a series, anticipatory by 3 sec). The skin response is also prominent on the scalp, and a K wave is present in all EEG leads. In addition, the entire character of the EEG changes.

a frequent electroencephalographic concomitant of the galvanic skin reflex. For their identification, it was necessary to make a distinction on the one hand from movement artifacts, especially accompanying blinking, and on the

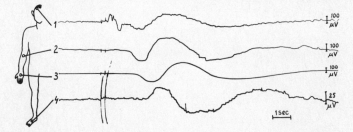

Fig. 6. Increase of the latency of the galvanic skin reflex with increasing distance of the lead from the head. A K wave is also present in the recording from the head.

other hand from a rudimentary galvanic skin reflex on
the scalp. From Figs. 4, 5, 9, 10, 14, 15, and 16, in which

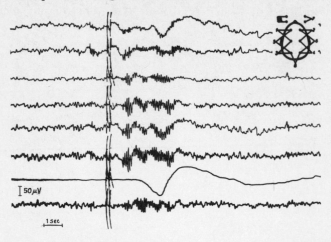

Fig. 7. Unconditioned reflex. The skin response is also present in the first of the scalp leads, and three separate spindles of alpha rhythm are evident in the EEG.

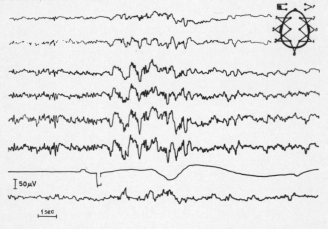

Fig. 8. Isolated unconditioned reflex to an auditory stimulus. The skin response is present and slow waves are evident in the EEG in all channels.

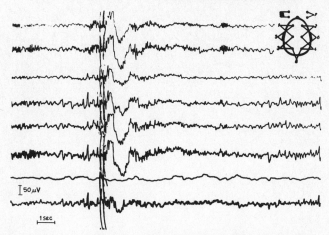

Fig. 9. Unconditioned reflex. A K complex is present in all channels of the EEG; the skin response is completely absent.

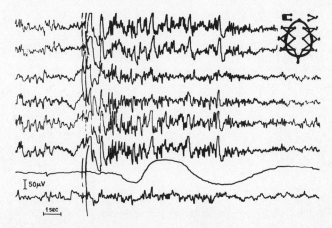

Fig. 10. Unconditioned reflex. A skin response is present, and in the EEG there is a salvo of K waves in all channels.

examples of K waves are shown, it is evident that the differentiation from blink artifact is not difficult. The latter consists of a wave that is always monophasic and of shorter duration (Figs. 13, 18, and 22).

Differentiation from the galvanic skin reflex of the scalp is for the most part possible at a glance, from the

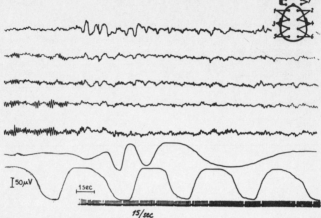

Fig. 11. Unconditioned reflex, occurring exceptionally, to flicker. The skin response oscillates, there is a blocking of alpha rhythm, and respiration is accelerated (8th channel).

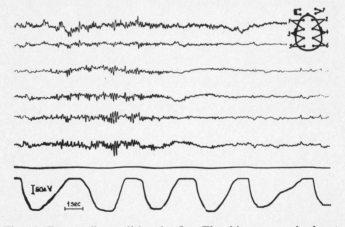

Fig. 12. Temporally conditioned reflex. The skin response is absent; only a blocking of the alpha rhythm in the EEG and an acceleration of respiration (8th channel) are evident. Such a temporally conditioned reflex was not included in the numerical evaluation of the results because of its incompleteness.

morphology of the two phenomena. In doubtful cases they can be reliably differentiated by their latency: That for the K wave is always shorter than that for the galvanic skin reflex, although the latter is dependent on the site of recording, being the shortest on the head. From comparisons of the K wave and the galvanic skin reflex from various places on the body (Fig. 6) it was ascertained that the latency of the galvanic skin reflex in the hand is around 1.5 sec, on the foot between 2 and 2.5 sec, whereas on the head it is around 1 sec. The K wave, however, has a latency that is always less than 0.2 sec (Roth et al. [1956] mentioned a maximum of 170 msec). Therefore, in those cases where the response appeared without any stimulus, thus making it impossible to measure latencies (i.e., in temporally conditioned reflexes, see below), the K wave always preceded the galvanic skin reflex on the scalp.

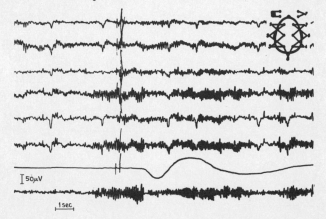

Fig. 13. Unconditioned reflex. There is a skin response but no change in the EEG. Six blink artifacts (monophasic waves downward pointed) are present.

The K wave is found typically in sleep; in the waking state it is present only in some subjects. Roth et al. (1956)

mentioned finding it in 20 per cent of waking subjects, Gastaut (1953) in 25, and Bancaud et al. (1953) in 50 per cent. In my experimental series, it was found in 11 of 15 subjects, that is, in 73 per cent (cf. Table 1),

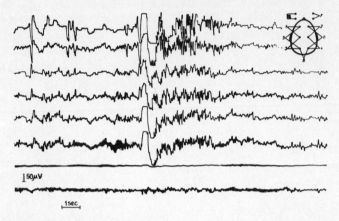

Fig. 14. Temporally conditioned reflex (second in a series, delayed by 2 sec). AK complex is present in all EEG leads, but because the skin response was absent, the reflex was not evaluated.

although the subjects were not in the sleeping state. Of course in some cases it was a question of the initial stage of falling asleep, or drowsiness, but in addition the more frequent appearance of K waves in waking subjects in my experiments could also be attributed to the circumstance that I worked with them repeatedly for long periods of time, so that in some the K wave was exhibited beyond the point of being inconstant. I judged the stage of wakefulness or sleep according to the usual EEG criteria (see, for example, Kayser 1949, Roth 1957), according to the responsivity of the subjects, on how they responded to questions, and according to the occurrence of the conditioned reflexes.

3. In the third group of EEG changes which accompanied the galvanic skin reflex, we consider changes of

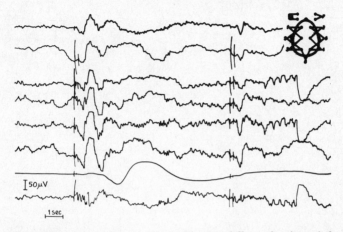

Fig. 15. Two unconditioned reflexes. The first falls at the time of the appearance of a temporally conditioned reflex, and both the skin response and the K wave are present in all EEG leads. The second falls in the interval between the temporal reflexes; here only the K wave in the EEG appears, the skin response being absent.

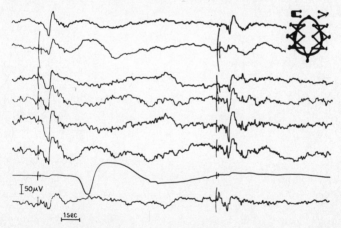

Fig. 16. Another instance of inhibition of the unconditioned reflex, from the same experiment as that of Fig. 15.

the entire character of the EEG; this manifested rather a variety of forms which had common features, however, of generalized changes that appeared bilaterally and simultaneously over the entire head. These included: the blocking of the alpha rhythm (a manifestation of increased

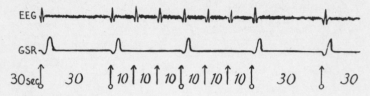

Fig. 17. Schematic representation of inhibition of an unconditioned galvanic skin reflex in the intervals between the appearance of temporally conditioned reflexes. In response to the stimuli, which are denoted by arrows, there is always an EEG response (a K wave), whereas the skin response (GSR) appears only every 40 sec. (Redrawn from the actual recording.)

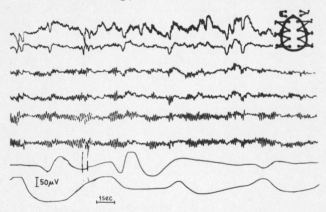

Fig. 18. Two reflexes appearing as isolated skin responses, without change in the EEG. Two seconds before the stimulus for the unconditioned reflex there is an anticipatory temporally conditioned reflex. A blink artifact appears 7 times (waves pointed downward). The 8th channel is the pneumogram.

attention), as in Figs. 11 and 12; at other times, in contrast, the reappearance of alpha waves (during the transition from drowsiness to wakefulness), as in Fig. 7;

at still other times, the appearance of regular slow waves, as in Fig. 8, or spindles of activity of 14 per sec, as in

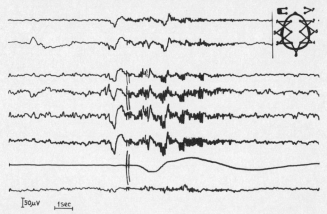

Fig. 19. Anticipatory conditioned reflex to time, presenting as an isolated K wave in all EEG leads; this appears just before an unconditioned reflex that presents as a skin response and as a change in the entire character of the EEG.

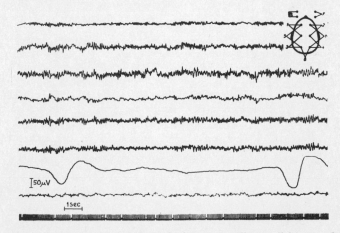

Fig. 20. Decrease of the interval of a temporally conditioned reflex from 30 sec to 13 sec as a result of flicker of a frequency of 14 per sec. Both reflexes appear only as skin responses without change in the EEG. The imposed rhythm is apparent in all EEG leads.

Fig. 3 (as a manifestation of a shift from a deeper to a lighter stage of sleep). The appearance of the K complex (Figs. 9 and 14), or a whole series of K waves (Fig. 10), which was also observed, properly belong to the second and third groups of EEG changes.

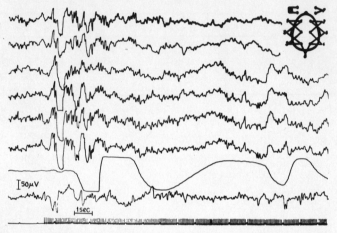

Fig. 21. Decrease of the interval of a temporally conditioned reflex from 30 sec to 13 sec at the onset of flicker of a frequency of 14 per sec. Both reflexes appear as a skin response and as a K wave in all EEG leads; the first reflex is more prominent, since it is unconditioned (the onset of flicker being the stimulus). No imposed rhythm (photic driving) is apparent.

In view of the well-known individual variability, fluctuation, and gradual weakening of the galvanic skin reflex (Gildemeister 1928, Goadby and Goadby 1949, Davis 1934), it is not surprising that the EEG changes accompanying the galvanic skin reflex did not always occur simultaneously, but for the most part only some of them alone or in various combinations, as is apparent, for example, in Figs. 9, 13, and 20. We note at once that the same holds for unconditioned as well as conditioned reflexes to time; all of the EEG changes described were found in both (see below).

Dissociation of the Galvanic Skin Reflex and the Accompanying EEG Changes (Holubář 1958b, 1959a)

In view of the fact that under certain circumstances the galvanic skin reflex becomes extinguished and sometimes simultaneously the skin sensitivity decreases (Holubář 1957), it seemed important to observe different instances of this dissociation more closely, from the standpoint of the change of skin sensitivity. The simultaneous bilateral and generalized appearance over the entire head of waves was a common characteristic of the EEG changes, whether falling in the second or third group according to the distinctions described. It is possible at this point to distinguish four basic types of responses to nociceptive stimuli:

1. Presence of both the skin and the EEG responses; this was the most typical and the most frequent type.

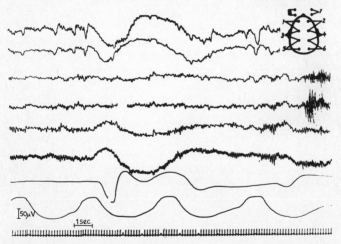

Fig. 22. Decrease of the interval of a temporally conditioned reflex from 30 sec to 9 sec as a result of flicker of a frequency of 7 per sec. Both reflexes appear as a skin response that is also evident on the scalp. No imposed rhythm is apparent in the EEG. Several blink artifacts are evident in the frontal leads. Channel 8 is the pneumogram.

2. Absence of both responses, even when the shocks were quite appreciable, to the point of being almost painful. In general, this type occurred infrequently in subjects who otherwise responded normally, and in general it was possible later to ascertain that they had been suffering from infections of the upper respiratory tract, incipient influenza with increased temperature, or were convalescing from influenza. It was thus a question of a nonspecific disorder of vegetative equilibrium. Even in the preliminary experiments I had noted an absence of reflex changes of the electrical skin resistance in similar circumstances.

3. The absence of any kind of EEG response in the presence of a fully developed skin response (Figs. 13, 18, and 20). This was sometimes noted as a transient phenomenon at the very beginning of a session, when the subject had not yet become sufficiently at ease. One can perhaps presume that a galvanic skin reflex (the skin sensitivity being normal) in such a case occurred predominantly at the lower levels of the central nervous system, without any generalized response in the cerebral cortex, as though the diffuse projection of afferent impulses to the cerebral cortex (Holubář 1957) were suppressed, or rather overshadowed, by activity of the waking state.

4. Regular appearance of a generalized cortical response, as the subject progresses to a relaxed state, as already described. In the initial stages of falling asleep, i.e., during a certain transitional stage between wakefulness and sleep, or drowsiness, a waning of the skin response to the point of complete disappearance could be noted, so that ultimately only an EEG response appeared to each stimulus (Fig. 9). The isolated K wave or K complex, appearing generalized over the entire head but with simultaneous complete absence of the galvanic skin reflex, was accompanied by complete or almost complete loss of skin sensitivity, as the subjects indicated during subsequent questioning, although the stimuli remained

physically unchanged. The fact that it was not a question of sleep was substantiated by the following criteria: by the responsivity of the subject, who reacted immediately and appropriately to questions, by the form of the EEG, and finally, by the occurrence of temporally conditioned reflexes in this state.

A similar case of dissociation of skin and EEG responses (absence of galvanic skin reflex in the presence of a K wave in all EEG leads) was noted to normal nociceptive stimuli in the time interval between temporally conditioned reflexes, as will be discussed in the following.

It is well known that the galvanic skin reflex gradually diminishes upon repeated presentations (Gildemeister 1928, Goadby and Goadby 1949, Davis 1934). This ready extinguishability appeared in my experiments under various other conditions: The galvanic reflex was completely absent in disorders of vegetative equilibrium and at the beginning of mild illnesses; it also disappeared during the process of falling asleep and during drowsiness; and it became extinguished during the conditioned inhibition in the intervals between the appearances of the temporally conditioned reflexes. The same mechanism could certainly not be relevant to all of these instances, as is indicated by the different behavior of the EEG and the variations in skin sensitivity in the individual cases. The correlation of changes of skin sensitivity with the galvanic skin reflex in my experiments can perhaps best be summarized by stating that the galvanic skin reflex disappears with the depression of skin sensitivity, but not conversely.

It has not so far been possible, evidently, to formulate a clear picture of the mechanism for the variation in intensity of the galvanic skin reflex, the variation of skin sensitivity to nociceptive stimuli, and the variation of the accompanying EEG changes. Wang et al. (1956, 1956a, b) described in cats an inhibitory center in the bulbar part of the reticular formation for the galvanic skin reflex, which could be experimentally activated from higher levels of the central nervous system, such as for example from the frontal cerebral cortex. As a rule, however, the brain stem facilitates the galvanic skin reflex. There is no doubt that one of the descending effects of the reticular formation is the control of the galvanic skin reflex.

There are in addition many reports on the control by the reticular formation of attention to specific perceptions, in part by means of a primary effect on the sensitivity of receptors (Granit 1955a, Livingston 1958, Holubář 1957). The first of these was found in proprioception, in the elucidation of the role of the gamma innervation in the cat (Granit and Kaada 1952, Granit et al. 1952, Granit and Holmgren 1955, Granit et al. 1955, see also Drechsler and Škorpil 1958), then for vision (Granit 1955b, Hernández-Péon, Scherrer, and Velasco 1956), for the auditory system (Galambos 1956, Hernández-Péon, Scherrer, and Jouvet 1956, Hernández-Péon, Jouvet, and Scherrer 1957), probably also for somatic sensitivity (Hernández-Péon and Scherrer 1955, Scherrer and Hernández-Péon 1955, Hernández-Péon, Scherrer, and Velasco 1956), even in man (Hagbarth and Höjeberg 1957).

Although the significance of all of these factors for man cannot at this time be fully evaluated, I believe that it can be concluded provisionally from my findings that there exists in man a state of inhibition in which the reticular formation transmits to the cerebral cortex a diffuse stream of nonspecific afferent impulses and at the same time inhibits both the galvanic skin reflex and the nociceptive sensitivity of the skin, sometimes by means of an intervention at higher levels.

Temporally Conditioned Reflexes (Holubář 1958a, b, 1959a)

As was the case for reflex changes of electrical resistance of the skin, establishing the usual conditioned reflexes for the reflex changes of skin potential and the accompanying EEG changes (which in the second case were examined only for orientation) was found to be comparatively very easy, whereas establishing temporally conditioned reflexes was laborious. For determining whether a temporally conditioned reflex was established, strict criteria were laid down: 1. There must be a skin response; an EEG response alone was insufficient. 2. The response must appear at least three times in succession without the presentation of stimuli at the appropriate interval with sufficient accuracy, that is, with a maximum admissible deviation in time of ±10 per cent. 3. The response must not be interfered with by spontaneous responses which would affect the evaluation of the recordings. Even so, the establishment of a temporally conditioned reflex was successful in 14 out of 15 subjects, i.e., in 93 per cent (Table 1). To accomplish this, however, a large number of reinforcements (repetitions) were necessary, 100 or even a few hundred, as well as several sessions. Thus it was not possible to establish a temporally conditioned reflex (according to the criteria given) in a single or in two sessions, even if they followed one another directly. The requirement of three responses in succession was often surpassed, as many as 11 temporally conditioned reflexes being attained in succession without the presentation of stimulus.

At a comparatively early stage in an experiment, in the course of establishment of temporally conditioned reflexes, I frequently observed in a series of unconditioned reflexes an occasional response before the full interval had passed (Figs. 18 and 19). It was probably a matter of an anticipatory temporally conditioned reflex, so that if the strict criteria were not observed, the establishment of a temporally conditioned reflex would even appear to be relatively easy. With gradual establishment, however, the intervals of the temporally conditioned reflex became more precise, so that established reflexes showed considerable accuracy, with a maximum variation of some 10 per cent. To be sure, among the numerical values of the intervals which were used for quantitative evaluation, even much greater deviations occurring in the case of still unstable reflexes could not be discarded, so as to avoid arbitrariness (Figs. 23, 25, 27, and 29).

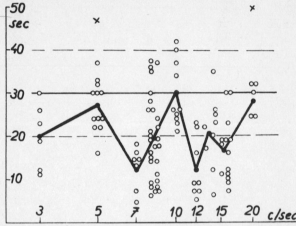

Fig. 23. Dependence of the duration of the intervals of temporally conditioned reflexes (vertical axis) on the frequency of flicker (horizontal axis, logarithmic scale). Open circles denote individually measured values; filled circles indicate arithmetic means, which are connected by the broken line only for clarity. The x's denote sporadic instances of prolongation of intervals beyond 30 sec, which were not included in the means. The solid horizontal line denotes the average of intervals without flicker, the dashed lines above and below it demarcate the entire dispersion of intervals without flicker, that is, the extreme values. Subject PR.

Whether or not these anticipatory responses represent the beginning of a temporally conditioned reflex, it seems to me that yet another significance can be attributed to them. In the Introduction it was mentioned (p. 9) that two explanations exist for the mechanism of temporally conditioned reflexes: according to Pavlov, a coupling to a suitable internal rhythm of the organism, and according to Hull, coupling to the end or consequence of a preceding response. If in the establishment of temporally conditioned reflexes the anticipatory responses appear without interfering with the establishment, such a finding would

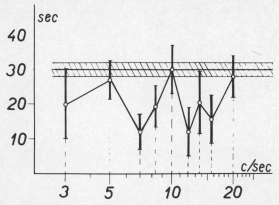

Fig. 24. Circles denote the arithmetic means, the vertical bars indicate 3 times the mean error of the arithmetic averages (for formula see Table 2). The middle horizontal line denotes the average of intervals without flicker, the shaded strip again indicates 3 times the mean error of the arithmetic average of the intervals without flicker. In this way, the statistical significance of the differences is represented. Subject PR.

attest rather to the validity of Pavlov's explanation, for if the theory of Hull were correct, the anticipatory responses would disturb the formation of a rhythm for the next temporally conditioned reflex. Some responses would thus occur at a different phase of the preceding ones. Of course it must be admitted that the argument is rather a qualitative one, for in my experiments no quantitative demonstration was carried out to show that the elaboration of a temporally conditioned reflex was at all disturbed by a preceding response.

So far as a picture of the temporally conditioned reflex is concerned, I have found all of the changes described for the unconditioned reflex are also manifestations of the conditioned reflex, individually or in various combinations. The most interesting was the conditioned reflex occurrence of a K wave (Figs. 5, 14, and 19) found in all

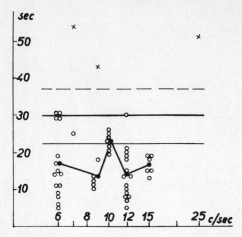

Fig. 25. For explanation see Fig. 23. Subject VV.

of the 11 subjects in whom K waves were at all apparent (Table 1). Because of the nonspecificity of the K wave

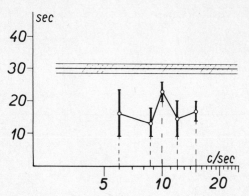

Fig. 26. For explanation see Fig. 24. Subject VV.

(see p. 47), it was not readily possible to demonstrate its conditioned reflex appearance by the usual means of conditioning, for its appearance to any kind of indifferent

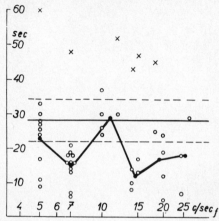

Fig. 27. For explanation see Fig. 23. Subject KL.

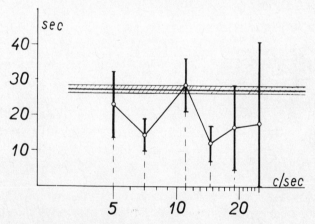

Fig. 28. For explanation see Fig. 24. Subject KL.

stimulus could always be interpreted as a possible unconditioned reflex. Only as a temporally conditioned reflex, where no external stimulus of any kind is presented, is it possible to consider a conditioned reflex occurrence of the K waves as conclusive.

The K wave, for the most part, is regarded as similar or identical to the so-called secondary cortical response (Forbes and Morison 1939). The latter is a generalized electrical response of the cerebral cortex, arising non-specifically from stimulation of any kind of receptor, as a result of arrival of impulses via the reticular formation, i.e., as nonspecific afferent impulses to the entire cortex (Holubář 1957). Gastaut (1953, 1954) emphasizes the predominant occurrence of the K wave at the vertex, and holds that this wave is the manifestation of the arrival of nonspecific impulses from the brain stem to the cortex of the limbic system. Roth et al. (1956) considered the K wave as the manifestation of simultaneous activation of different parts of the cerebral cortex by some diffuse projection system. Larsson (1956) demonstrated a mesencephalic origin of the K wave. The view, held by Oswald (1959), that the K wave is a movement artifact, is an isolated and surely incorrect one. It seems then, that for the origin of the K wave the nonspecific afferent system, originating in the brain stem, can be considered, and the finding of the possibility of conditioning the K wave can be interpreted as further evidence supporting the hypothesis of the dominance of the origin of temporary connections in the reticular formation (Yoshii et al. 1956, 1957, Gastaut 1958) or at least of a primary contribution of the subcortex to the formation of temporary connections.

Mention has already been made of the fact that a particular manifestation of inhibition was observed in the time interval between temporally conditioned reflexes. After further elaboration of the temporally conditioned reflex by means of regular applications of stimuli at intervals of 30 sec, both the skin and the EEG responses (in the present case, the K wave) could be obtained by presentation of the stimulus at the proper moment. At intervening times, however, for example at 10 or 20 sec, the same stimulus resulted in practically the same EEG response, but the skin response was absent (Figs. 15 and 16). In one case, this phenomenon was demonstrated four times in succession (Fig. 17). This phenomenon cannot be due to some kind of refractoriness, for the galvanic skin reflex can be repeated in man even after 3 sec (Pampiglione 1952). Rather, it is a process in which the unconditioned reflex or its principal component becomes extinguished during the establishment of a conditioned inhibition. This indicates the great strength of inhibition in the time interval between successive temporally conditioned reflexes. Asratian (1953) describes only a weakening of unconditioned reflexes (defensive movements) with strong positive conditioned reflexes, repeated rapidly in succession in intervals of 5 to 30 sec. The inhibition of conditioned reflexes during extinction of another reflex (the so-called secondary extinction) is of course well known (Ellson 1938, Hill and Calvin 1939, Youtz 1939), even for a delayed conditioned reflex (Rodnick 1937).

The Specific Effect of Flicker on the Intervals of Temporally Conditioned Reflexes

A preliminary communication concerning these results has been published previously (Holubář 1959c). Unfortunately, a frequent effect of flicker on the temporally conditioned reflex was that the reflex did not appear. This property of temporally conditioned reflexes (that of ready inhibition by an external stimulus) was well known from the earliest studies (Stukova 1914). Its nonspecific effect disqualified many subjects from the final series of data, so that only half of the subjects gave interpretable results; among those, the data were so small for 3 subjects as to be inconclusive. Remaining then for the final evalua-

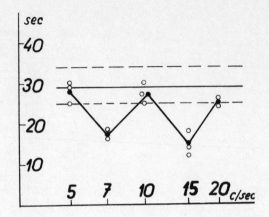

Fig. 29. For explanation see Fig. 23. Subject UV.

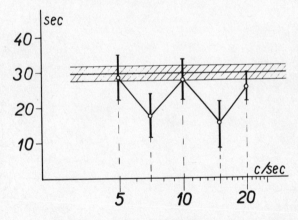

Fig. 30. For explanation see Fig. 24. Subject UV.

tion there were only 4 subjects (Table 1), but for these the data were altogether consistent and not a single result for any of the subjects was at variance with the rest. The number of intervals evaluated for the temporally conditioned reflexes was further limited to those cases that showed a skin response, whereas cases of isolated EEG

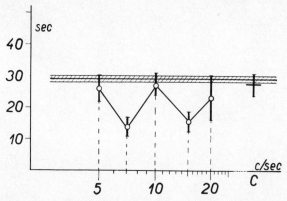

Fig. 31. For explanation see Fig. 24. Averaged values for the 4 subjects. The average and the dispersion denoted by C refers to the control experiments with continuous light.

responses were not evaluated, since these could be considered to be defective reflexes. Nevertheless, the entire number of evaluated intervals was 258 (Table 2).

The data are shown in graphs (Figs. 23 to 30) and in Table 2. The former show the intervals of the temporally conditioned reflexes for various frequencies of flicker, in comparison with the same intervals without flicker (i.e., as controls) for individual subjects; finally, in Fig. 31, the results from all of the subjects collectively are summarized. For each subject there are shown both the individual distribution of intervals for flicker, in comparison with the over-all dispersion of values of intervals for the controls (Figs. 23, 25, 27, and 29), as well as the dispersion of values during flicker and for the controls, expressed as three times the mean error of the arithmetic mean, so as to indicate the statistical significance of the

Table 2. Effect of Flicker on the Intervals Between Temporally Conditioned Reflexes. (Averages and Dispersions of the Results for Subjects PR, VV, KL, UV)

Frequency of flicker (c/sec)	Number of measurements n	Average interval m (sec)	Mean error[1] s_m (sec)	Significance of the differences[2] P
without flicker	140	29	0.31	
5	24	26	1.44	0.04
7	21	14	1.04	$<10^{-9}$
10	26	27.5	1.21	0.23
14–15	28	15.5	1.10	$<10^{-9}$
20	12	23	2.39	0.01
continuous light	7	27	1.23	0.12
	258 Total			

[1] Mean error of the arithmetic average computed according to the formula $s_m = \sqrt{\Sigma(x^2)/(n-1)}$.
[2] The significance of the differences are with respect to the value for no flicker; P is the probability computed by means of the t test.

differences (Figs. 24, 26, 28, and 30). The decrease of the intervals of the conditioned reflexes to time, due to flicker, is shown in Figs. 20, 21, and 22.

All of the data similarly showed a significant decrease of the intervals of the temporally conditioned reflexes under the influence of flicker at certain frequencies, especially at 7 and 14 to 15 per sec, for which the decrease amounted on the average to approximately one half, i.e., 14 and 15.5 sec instead of 29 sec. In contrast, for a frequency of 10 per sec, the intervals remain almost entirely unchanged, and for frequencies of 5 and also 20 per sec, the effect (a decrease) was insignificant, on the average. This striking dependence of the effect of flicker on frequency, a maximum of the effect for frequencies of 7 and 14 to 15 per sec, and a minimum or no effect for frequencies of 5, 10, and 20 per sec, could be termed the "rule of octaves."

This specific dependence of the effect on the frequency of the flicker itself shows that there can be no question of a nonspecific disinhibition of the influence of external stimuli that were particularly effective. Nevertheless, according to plan, I examined the effect of continuous light as another kind of control. It was found that it had practically no effect on the intervals of the temporally conditioned reflexes under our experimental conditions (Table 2 and Fig. 31); this additionally underscores the specificity of the "rule of octaves."

Of course it cannot be excluded that the effect of flicker on the interval of temporally conditioned reflexes may involve some small nonspecific component, i.e., a small decrease of the interval, with no dependence on frequency. Such an impression is suggested by the data for subject VV, for whom the entire graph appears to be shifted downward (Figs. 25 and 26).

Sporadically, for various frequencies, instances arose of an apparent prolongation of intervals beyond 40 sec (twice for subject PR, 3 times for subject VV, and 6 times for subject KL, for a total of 11 occurrences). In the graphs (Figs. 23, 25, and 27) such values are denoted by crosses. In view of their rarity, I do not consider that they result from an actual prolongation of intervals but rather from a doubling of shortened or unchanged intervals, arising in such a manner that in a series of temporally conditioned reflexes one dropped out, since the extinction of a reflex was a frequent phenomenon during flicker.

As follows from the graphs of Figs. 24, 26, 28, 30, and 31, and from Table 2, the statistical significance of the results is high. Still higher is the significance of the difference between individual frequencies (for example, between 7 and 10 per sec, or between 10 and 14 per sec) considered as a pair; this method can perhaps be justified in view of the small nonspecific component of the effect. Nor does the apparent prolongation of values of the interval jeopardize the statistical significance of the results by being included in the calculation.

In view of the fact that the alpha rhythm in the subjects studied fell, on the average, between 10 and 10.75 (see Table 1), one can speak of the alpha frequency and its multiples, with respect to the frequencies of flicker employed, i.e., 5, 10, and 20 per sec, with a sufficient degree of approximation.

Discussion of the Results from the Standpoint of the Working Hypothesis

The results of the experiments with the imposition of flicker on the intervals between the temporally conditioned reflexes fulfilled the prediction of the working hypothesis, and in addition introduced some unexpected findings. Fulfillment of the prediction consisted in that a prominent specific effect on the intervals of the temporally conditioned reflexes by the flicker was found, thus underscoring the possibility of a significant role for brain rhythms in the sense of time, and hence of the possibility of locating the pendulum of the biological clock in the brain of man. In addition, the findings speak strongly in favor of the Pavlovian theory with respect to the mechanism of temporally conditioned reflexes (see p. 9). Further information, which can perhaps be included under the term "rule of octaves" and which could cast further light on the mechanism of time sense, can be summarized in the following points:

1. Frequencies of flicker that do not coincide with the alpha rhythm but are themselves harmonically related, that is, 7 and 14 to 15 per sec, are maximally effective.

DISCUSSION OF RESULTS

2. Both of these frequencies, of which one is lower and the other higher than the alpha rhythm, result in a significant, and on the average almost identical, decrease of the intervals between the conditioned reflexes to time.

3. Flicker coinciding with the frequency of the alpha rhythm is without effect.

4. Flicker of frequency of half (5) and double (20) the alpha rhythm has a negligible effect on the whole; for simplicity these frequencies can also be considered as ineffective.

Concerning 1: Although the most easily imposable rhythm on the EEG is one of 10 per sec, i.e., as a rule the highest voltage is attained for it, the imposition of this rhythm has the least striking effect because it is already present as the spontaneous alpha rhythm. In contrast, rhythms of 7 and 14 to 15 per sec result in the greatest change in the character of the EEG as a whole and are very conspicuous as imposed rhythms. Their maximal effect on the sense of time may be related to this point.

Concerning 2: Two interpretations are possible. Either with the imposition of a new rhythm on the brain the usual time-measuring mechanism was displaced and a completely different mechanism, unknown to us, became operative and conceivably resulted in a constant error, however the original rhythm was displaced. Or, the imposed rhythm became the basis for measurement of time and resulted in a constant error, independent of whether the flicker were 7 or 14 per sec, such that in both cases, the frequency of imposed rhythm either was not at all or largely not distinguished (as, for example 7 per sec, from its characteristic harmonic of 14 per sec). It is also well known that imposed rhythms frequently appear as a multiple or half of the frequency of the flicker. The determination could perhaps be made by carrying out frequency analysis of the EEG recording, which would enable an exact approportioning of the separate frequencies in the EEG during the experiments.

Concerning 3 and 4: These facts strongly impel the supposition, also supported by the preceding points, that the time-measuring rhythm sought for could be the alpha rhythm itself. The latter is, after all, the most prominent autonomous rhythm of the human brain. Of course, to deal with the question of *how*, explicitly, the experimental material obtained is very inconclusive; it is, rather, only suggestive. In addition, many facts are known about the alpha rhythm itself, but very few of these are of a basic nature.

Many reports in favor of the significance of the alpha rhythm or, more generally, the electrical rhythms of the brain, in the mechanism of the sense of time could be mentioned from the literature. Thus, for example, the beginning of the manifestations of the sense of time at 4 years of age and the achievement of its full development by puberty (Binet and Simon 1916) are very well in accord with the age for regularity of brain rhythms (Schütz et al. 1951, Hill and Parr 1950). It is also relevant that both the sense of time and the brain rhythms are dependent on body temperature (Hoagland 1935, 1936). These and similar parallels, however, prove little, or rather disprove nothing, concerning the role of brain rhythms in the sense of time, for ontogenesis, change with temperature, etc., concern a great number of different functions of the organism that develop or change in parallel; this still does not prove a primary causal relationship between them, but rather a common cause in the background.

The origin of the alpha rhythm has not so far been satisfactorily explained. There are two groups of theories that attempt an explanation (see, for example, Brazier

1951, Adrian 1947): Either there is the production of a rhythmic oscillation as an inherent property of neurons, and the alpha rhythm results accordingly whenever the oscillations of a large number of neurons, primarily of the cortex, become synchronized. Or, emphasis is placed on the more complex combinations of different parts of the brain, for which the interneuronal connections of brain structure would be essential.

In support of the first view, it can be mentioned that rhythmic activity arises also in the isolated cortex (Bremer 1949) and that the alpha rhythm and its sensitivity to light already occur in as simple a nervous system as that of the water beetle (Adrian 1931), which indicates the manifestation of a general characteristic of cellular masses rather than the significance of the particular anatomical organization of the cells. Synchronization of activity of individual neurons placed in a conducting medium, whether implemented primarily chemically or electrically, have been reported, for example, in well-known work (Libet and Gerard 1939) in which the anatomical division of the frog brain, by means of cuts, did not prevent the spread of electrical waves from one part to another, or in a study (Arvanitaki 1942) with isolated axons of invertebrates, the oscillations of which became synchronized by mere contact.

Among the second group of theories, perhaps the best known is the interpretation of the alpha rhythm as the circulation of impulses between the cerebral cortex and the thalamus (Dusser de Barenne and McCulloch 1941). The significance of the thalamus and its connections with the cortex for the origin of the alpha rhythm was emphasized in the work of Dempsey and Morison (1943). More recently, reports have increased concerning the existence in the brain of multiple generators of alpha activity which are capable of being synchronized, whereupon the alpha rhythm becomes manifest, or, the phases become dispersed, whereupon it ostensibly disappears (Bekkering et al. 1957, Walsh 1958).

It would seem that, from the information cited and from many others that concern the origin of the alpha rhythm, even though so far no consistent and complete concept has been offered, it is possible to select two facts for our purposes, i.e., as support for the identification of the alpha rhythm with the time-measuring mechanism. On the one hand, there is the significance of the thalamus and thalamo-cortical connections, both for the origin of the alpha rhythm and for the sense of time (see Introduction, p. 25). On the other hand, there is the important circumstance that alpha activity of separate generators or perhaps of neurons is not abolished in periods when no alpha rhythm is manifest in the EEG, but persists although not outwardly apparent, because the activity of the majority of generators, or nerve cells, is not synchronized. By this reasoning, the objection can be refuted of how the sense of time could function during periods of absence of the alpha rhythm in the EEG, i.e., for a great part of life. On the other hand, the very circumstance that is favorable for manifestation of the alpha rhythm, i.e., undisturbed quiet, is at the same time favorable for accurately maintaining the sense of time.

The seemingly contradictory results of prolongation of time intervals by flicker, in the subjective method, were discussed under that method (p. 41).

In view of the fact that other rhythms, especially the pulse and respiration, have previously been suggested as possible time-measuring mechanisms (see Introduction, p. 22) and in view of reports in the literature on a relation between the rhythm of respiration and EEG rhythms in rabbits (Gurevich 1948, 1949), it was still necessary to resolve the question of whether, and to what degree, flicker may influence the rhythm of breathing or the

DISCUSSION OF RESULTS 79

pulse rate, and whether then the effect of flicker on the sense of time could not be complicated by such relationships (Holubář 1959d).

Experiments on this point carried out in human subjects were entirely negative; neither the pulse rate nor respiration were at all influenced by flicker (except for a nonspecific brief acceleration of respiration which sometimes occurred at the beginning of flicker, as in Fig. 11). Nor was a synchronization of spindles (of imposed rhythms, of the alpha rhythm, or during sleep) with breathing ever found in man. Perhaps in this instance, it is appropriate to remark that in temporally conditioned reflexes, when respiration was also recorded, an acceleration of respiration was frequently noted which accompanied the entire complex of the remaining changes (Fig. 11 for the unconditioned, and Fig. 12 for the conditioned reflex). Thus it can be concluded that the effect of flicker on the sense of time in man is not complicated by any secondary effect either on respiration or on the pulse. The absence of an effect on the pulse rate is not surprising, for the automatism of the heart hardly allows the possibility,

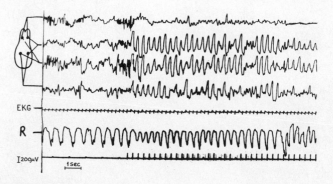

Fig. 32. The imposed rhythm of the EEG of a rabbit during flicker with a frequency of 3 per sec displaces the respiratory rhythm (pneumogram R) to the same frequency.

for example, of some kind of synchronization of rhythm by means of a neural control. As far as respiration is concerned, a correlation between imposed EEG rhythms and respiration was found in rabbits.

Here the EKG and pneumogram were recorded simultaneously with the EEG by means of screw electrodes placed in the skull (Holubář 1959d) under local anesthesia. During flicker (to both eyes) of different frequencies, imposed rhythms appeared in the EEG for a wide range of frequencies. For sufficiently low frequencies (for example, around 3 per sec) respiration often assumed the rhythm of flashing; in this case all three rhythms were synchronized for a prolonged interval of time (Fig. 32).

Thus it would seem that, in the rabbit, the imposed EEG rhythms propagate in the brain deeply, subcortically, to the respiratory center, which does not occur in man.

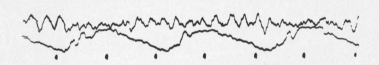

Fig. 33. Imposed rhythm in the EEG of a rabbit (upper curve) consisting of spindles which are synchronized with respiration (lower curve). Upper markings—flicker; lower markings—sec.

The reasons for this specific difference can perhaps be found in the fact that, in man respiration is much slower than in the rabbit, so that there is no commensurability between possible imposed EEG rhythms and the respiratory rhythm, as is the case in small animals. The circumstance of the smaller dimensions and simpler structures of the rabbit brain is assuredly relevant to the question of whether the propagation of the imposed rhythm through the brain occurs electrotonically and not by impulses, which is very probable (Libet and Gerard 1939).

In the rabbit it is obviously possible to induce appreciable hyperventilation by means of the flicker (Fig. 32). The thought suggests itself that such hyperventilation could be relevant to the origin of photogenic epileptic seizures, which we have described in rabbits (Holubář and Kohlik 1950).

In the rabbit, I have also noted the converse phenomenon, that of the imposition of the respiratory rhythm on the electrical activity of the cerebral cortex (Fig. 33). The voltage of the imposed EEG rhythm oscillates slowly and regularly for the most part, so that spindles of the imposed waves are apparent in the EEG. In cases of slow respiration, I observed synchronization of these spindles with respiration. It thus appears that the rhythm of the respiratory center is relayed to the cerebral cortex; this is apparently related to the lower degree of corticalization of functions in the rabbit, or to a higher status of the subcortex in the functional hierarchy.

CONCLUSION

Time is one of the fundamental quantities of physics. The perception of time is thus without doubt one of the most fundamental biological questions. The fact that we are constantly aware of the flow of time and of the order of all events signifies the very experiencing of life. Dreams are therefore imaginary (of course, in addition to other reasons), so that in them time is confused and, as it were, eliminated. The perception of time has for a long time attracted close investigation in psychology. The brief survey of the literature presented in the Introduction is intended for our problem only and cannot give a clear picture of the enormous effort of human ingenuity and work that has been exerted to discover the laws of the perception of time. A qualitative step in this effort was the discovery of temporally conditioned reflexes. This is one of the merits of Pavlov and his school which has not previously, I believe, been adequately appreciated. The entire investigation of the sense of time was removed from the field of materialistic physiology, although investigators clearly did not appreciate the fact; the two investigative trends, temporal conditioning and the psychological investigation of the sense of time, have heretofore proceeded separately.

The main part of this work is devoted to the study of temporally conditioned reflexes in man, by a complex method—the observation of the galvanic skin reflex simultaneously with several leads of the electroencephalogram, in some cases supplemented by additional functions. This method was chosen so that the results would be entirely objective, undistorted by intentional intervention by the subjects. The intervals between temporally conditioned reflexes were found to be very accurate, after appropriate establishment of the reflex; this permitted the specific effect of interrupted (flickering) light on the duration of the intervals to be shown. The specific effect is decisively dependent on the frequency of flicker employed. Under the influence of frequencies that did not coincide with that of the alpha rhythm, namely 7 and 14 to 15 per sec, a decrease of the intervals on the average to one half occurred. During flicker of a frequency of the alpha rhythm, 10 per sec, however, the intervals remained unchanged; they also remained practically unchanged on the average for frequencies of half (5 per sec) and double (20 per sec) the alpha frequency. These phenomena can be denoted collectively as "the rule of octaves."

These results led to the conclusion that rhythmic activity of the brain or a part of it could represent a fundamental reference rhythm which serves the organism in the measurement of time. Specifically, in man for times of the order of minutes it is the alpha rhythm, i.e., the most prominent human electroencephalographic rhythm of a frequency of about 10 per sec. The reasons that lead to this conclusion can be summarized as follows: It is a question of a rhythm that is autonomous, unchanging in the course of life (in adulthood), present even when it is not evident externally, and therefore very suitable for measurement of time. Moreover, flicker alters practically no function other than the EEG, for which however the effect is very prominent, the imposed rhythm rapidly spreading over the entire head.

Experiments with temporally conditioned reflexes are quite laborious and time-consuming, especially work with man which extends over months, and various technical difficulties are encountered. Further, the circumstance that flicker often completely prevents the manifestation of a temporally conditioned reflex seriously impeded the acquisition of the necessary quantity of data. These and other circumstances account for the fact that not all of the experimental approaches to the problem appropriate for the present state of research have been exhausted.

First of all, further studies should include widening the limits of the flicker frequency to include lower and higher values, since for the time being only frequencies were investigated in the range in which the imposition of the rhythms was the most successful, i.e., within two octaves, 5 to 20 per sec. So far only a few experiments have been carried out with frequencies of 3 and 25 per sec. A more closely spaced spectrum of frequencies would also be informative.

Further, there is a need to carry out analogous experiments with animals. Of course some other reflex would have to be chosen as an indicator, and the species of animal should also be selected with regard to the resting electrical activity of the brain. In the case of positive results, surgical or pharmacological measures could then perhaps be undertaken to modify the experimental results intentionally, and the localization of the time "center" could perhaps be determined more closely.

Since the intensity of the imposed rhythm in the course of flicker varies significantly, a correlation should be carried out between the degree of imposition of the new rhythm and the actual decrease of the interval in individual experiments. In investigating this relationship, I was not successful in finding an appropriate measure of the degree of imposition of the rhythms; a judgment of the presence or prominence of the imposed rhythm was very subjective, if not impossible. Frequency analysis of the EEG recordings and crosscorrelation of the EEG with the flicker would be appropriate as quantitative indications of the degree to which the external rhythm is imposed and of the degree to which the intrinsic rhythm is suppressed in individual cases.

These are some working suggestions drawn closely from the work carried out and appropriately complementary to it. In addition, there remain a series of more remote questions and problems that arise from our investigation, as for example experiments on the effect of flicker on the navigation of birds, the effect of flicker on the sense of time in man during sleep, and in general the possibility of affecting the measurements of longer time intervals by flicker, i.e., the problem of what is the longest time for which brain rhythms could serve as a time-measuring pendulum, as well as many other questions. The problem of the sense of time is still far from being resolved.

REFERENCES

Adrian, E. D.: Potential changes in the isolated nervous system of Dytiscus marginalis. J. Physiol. (Lond.) 72 : 132, 1931.

Adrian, E. D.: The physical background of perception. Oxford 1947.

Adrian, E. D., Matthews, B. H. C.: The Berger rhythm: potential changes from the occipital lobes of man. Brain 57 : 355, 1934.

Alekseev, M. A.: On the problem of neural mechanisms and interaction of two cortical signaling systems during conditioned rhythmical motor responses in man. Zh. Vyssh. Nerv. Deiat. 3 : 883, 1953. (In Russian)

Arvanitaki, A.: Effects evoked in an axon by the activity of a contiguous one. J. Neurophysiol. 5 : 89, 1942.

Asratian, E. A.: Physiology of the central nervous system. Moscow 1953. (In Russian)

von Baer, K.: Reden gehalten in wissenschaftlichen Versammlungen. St. Petersburg 1864. (cited by von Skramlik 1934 b).

Baiandurov, B. I.: Conditioned reflexes in birds. Tomsk 1937 (cited by Dmitriev and Kochigina 1955). (In Russian)

Bancaud, J., Bloch, V., Paillard, J.: Contribution EEG à l'étude des potentiels évoqués chez l'homme au niveau du vertex. Rev. neurol. (Paris) 89 : 399, 1953.

Bartley, S. H.: Temporal and spatial summation of extrinsic impulses with the intrinsic activity of the cortex. J. cell. comp. Physiol. 8 : 41, 1936.

Bartley, S. H.: Relation between cortical response to visual stimulation and changes in the alpha rhythm. J. exp. Psychol. 27 : 624, 1940.

Bartley, S. H., Bishop, G. H.: Cortical response to stimulation of the optic nerve in the rabbit. Amer. J. Physiol. 103 : 159, 1933.

Bekhterev, V. M.: On the reproductive and combinative responses during movements. Obozr. Psikhiat. Nevrol. eksp. Psikhol. 7 : 385, 1908. (In Russian)

Bekkering, D. H., Kuiper, J., Storm van Leeuwen, W.: Origin and spread of alpha rhythms. Acta physiol. pharmacol. Neerl 6 : 632, 1957.

Beritov, I. S.: Individually acquired activity of the central nervous system. Tbilisi 1932. (In Russian)

Benussi, V.: Psychologie der Zeitauffassung. Heidelberg 1913.

Binet, A., Simon, T.: The development of intelligence in children. Baltimore 1916.

Bolotina, O. P.: Conditioned motor reflexes to time in dogs. Trudy Inst. Fiziol. im. I. P. Pavlova 1 : 29, 1952a. (In Russian)

Bolotina, O. P.: Conditioned motor reflexes to time in monkeys. Trudy Inst. Fiziol. im. I. P. Pavlova 1 : 196, 1952b. (In Russian)

Bolotina, O. P.: Effect of bromide and caffeine on conditioned reflexes to time in dogs and monkeys. Trudy Inst. Fiziol. im. I. P. Pavlova 2 : 52, 1953. (In Russian)

Brazier, M. A. B.: The electrical activity of the nervous system. London 1951.

Bremer, F.: La nature des „ondes" cérébrales. EEG clin. Neurophysiol. 1 : 177, 1949.

Carrel, A.: Physiological time. Science 74 : 618, 1931.

Carrel, A.: Man, the unknown. Revised edition. New York 1939.

Čapek, D.: Physiology of aviators. Prague, 1953. (In Czech)

Cooper, L. F., Erickson, M. H.: Time distortion in hypnosis. Baltimore 1954.

Davis, H.: Space and time in the central nervous system. EEG clin. Neurophysiol. 8 : 185, 1956.

Davis, H., Davis, P. A., Loomis, A. L., Harvey, E. N., Hobart, G.: Changes in human brain potentials during the onset of sleep. Science 86 : 448, 1937.

REFERENCES

Davis, H., Davis, P. A., Loomis, A. L., Harvey, E. N., Hobart, G.: Human brain potentials during the onset of sleep. J. Neurophysiol. **1** : 24, 1938.

Davis, H. et al.: Analysis of the electrical response of the human brain to auditory stimulation during sleep: Amer. J. Physiol. **126** : 474, 1939a

Davis, H. et al.: A search for changes in direct-current potentials of the head during sleep. J. Neurophysiol. **2** : 129, 1939b.

Davis, H. et al.: Electrical reactions of human brain to auditory stimulation during sleep. J. Neurophysiol. **2** : 500, 1939c.

Davis, P. A.: Effects of acoustic stimuli on the waking human brain. J. Neurophysiol. **2** : 494, 1939a.

Davis, P. A.: The electrical response of the human brain to auditory stimuli. Amer. J. Physiol. **126** : 475, 1939b.

Davis, R. C.: Modification of the galvanic reflex by daily repetition of a stimulus. J. exp. Psychol. **17** : 504, 1934.

Dempsey, E. W., Morison, R. S.: The electrical activity of a thalamocortical relay system. Amer. J. Physiol. **138** : 283, 1943.

Deriabin, V. S.: Further studies on the physiology of time as a conditional stimulus of the salivary glands. Dissert. St. Petersburg, 1916. (In Russian)

Dmitriev, A. S., Kochigina, A. M.: The importance of time as a stimulus for conditioned reflex activity. Usp. Sovrem. Biol. **40** : 31, 1955. (In Russian)

Dobrovolskiĭ, V. M.: On trace feeding reflexes. Dissert. St. Petersburg, 1911. (In Russian)

Drechsler, B., Škorpil, V.: Recent findings on the innervation of striated muscles. Čs. Fysiol. **7** : 308, 1958. (In Czech)

Durup, G., Fessard, A.: L'électroencéphalogramme de l'homme. Observations psychophysiologiques relatives à l'action des stimuli visuels et auditifs. Année psychol. **36** : 1, 1935.

Dusser de Barenne, J. G., McCulloch, W. S.: Functional interdependence of sensory cortex and thalamus. J. Neurophysiol. **4** : 304, 1941.

Dvořák, J.: Some possible consequences of the theory of relativity for biological phenomena. Cs. Fysiol. **8** : 123, 1959. (In Czech)

Ehrenwald, H.: Versuche zur Zeitauffassung des Unbewussten. Arch. ges. Psychol. **45** : 144, 1923.

Ehrenwald, H.: Störung der Zeitauffassung, der räumlichen Orientierung usw. bei einem Hirnverletzten. Z. ges. Neurol. Psychiat. **132** : 516, 1931.

Einstein, A., Infeld, L.: The evolution of physics: the growth of ideas from early concept to relativity and quanta. New York, Simon and Schuster, 1938 (p. 189).
Ejner, M.: Experimentelle Studien über den Zeitsinn. Inaug.-Dissert. Dorpatt 1889.
Ellson, D. G.: Quantitative studies of the interaction of simple habits. I. Recovery from specific and generalized effects of extinction. J. exp. Psychol. 23 : 339, 1938.
Estel, V.: Neue Versuche über den Zeitsinn. Wundts Philos. Stud. 2 : 37, 1885.
Feokritova, J. P.: Time as a conditional stimulus for the salivary glands. Dissert. St. Petersburg, 1912. (In Russian)
Ferrari, G. C.: La psicologia degli scampati al terremoto de Messina. Riv. Psicol. 5 : 89, 1909.
Folk, G. E., Jr., Meltzer, M. R., Grindeland, R. E.: A mammalian activity rhythm independent of temperature. Nature (Lond.) 181 : 1598, 1958.
Forbes, A., Morison, B. R.: Cortical response to sensory stimulation under deep barbiturate narcosis. J. Neurophysiol. 2 : 112, 1939.
François, M.: Contribution à l'étude du sens du temps. Année psychol. 28 : 186, 1928.
Freeman, G. L., Sharp, L. H.: Muscular action potentials and the time-error function in lifted weight judgments. J. exp. Psychol. 29 : 23, 1941.
von Frisch, K., Lindauer, M.: Himmel und Erde in Konkurrenz bei der Orientierung der Bienen. Naturwiss. 41 : 245, 1954.
Frolov, Iu. P.: Physiological studies of I. P. Pavlov on time as a distinctive stimulus of the nervous system. Zh. Vyssh. Nerv. Deiat. 1 : 831, 1951. (In Russian)
Galambos, R.: Suppression of auditory nerve activity by stimulation of efferent fibers to cochlea. J. Neurophysiol. 19 : 424, 1956.
Gambarian, L. S.: On the question of conditioned defensive reflexes. Trudy Inst. Fiziol. im. Pavlova 1 : 73, 1952. (In Russian)
Gardner, W. A.: Influence of the thyroid gland on the consciousness of time. Amer. J. Psychol. 17 : 698, 1935.
Gastaut, H.: Etude électrographique chez l'homme et chez l'animal des décharges épileptiques dites „psychomotrices". Rev. neurol. (Paris) 88 : 310, 1953.
Gastaut, H.: In: Adrian, E. D. et al., ed.: Brain mechanisms and consciousness. Oxford 1954, (p. 261).

Gastaut, H.: The role of the reticular formation in establishing conditioned reactions. In: Jasper H. H. et al., ed: Reticular formation of the brain. Boston 1958. (pp 561–579).

Gellershteïn, S. G.: The sense of time and the rate of motor responses. Moscow, 1958. (In Russian)

Gildemeister, M.: Der galvanische Hautreflex. Bethe's Hdb. Physiol. **8** : 775, 1929.

Glass, R.: Kritisches und experimentelles über den Zeitsinn. Wundts Philos. Stud. **4** : 423, 1888.

Goadby, K. W., Goadby, H. K.: The nervous pathway of the psycho-galvanic reflex. J. Physiol. (Lond.) **109** : 177, 1949.

Gooddy, W.: Time and the nervous system. The brain as a clock. Lancet p. 1139, 1958.

Goudriaan, J. C.: Le rythme psychique dans ses rapports avec les fréquences cardiaque et respiratoire. Arch. Néerl. Physiol. **6** : 77, 1922.

Grabensberger, W.: Untersuchungen über das Zeitgedächtnis der Ameisen und Termiten. Z. vergl. Physiol **20** : 1, 1933.

Grabensberger, W.: Experimentelle Untersuchungen über das Zeitgedächtnis von Bienen und Wespen nach Verfütterung von Euchinin und Jodthyreoglobulin. Z. vergl. Physiol. **20** : 338, 1934a.

Grabensberger, W.: Der Einfluss von Salicylsäure, gelbem Phosphor und weissem Arsenik auf das Zeitgedächtnis der Ameisen. Z. vergl. Physiol. **20** : 501, 1934b.

Granit, R.: Receptors and sensory perception. New Haven 1955a.

Granit, R.: Centrifugal and antidromic effects on ganglion cells of retina. J. Neurophysiol. **18** : 388, 1955b.

Granit, R., Holmgren, B.: Two pathways from brain stem to gamma ventral horn cells. Acta physiol. Scand. **35** : 93, 1955.

Granit, R., Holmgren, B., Merton, P. A.: The two routes for excitation of muscle and their subservience to the cerebellum. J. Physiol. (Lond.) **130** : 213, 1955.

Granit, R., Job, C., Kaada, B. R.: Activation of muscle spindles in pinna reflex. Acta physiol. Scand. **27** : 161, 1952.

Granit, R., Kaada, B. R.: Influence of stimulation of central nervous structures on muscle spindles in cat. Acta physiol. Scand. **27** : 130, 1952.

Grassmück, A.: Mit welcher Sicherheit wird der Zeitwert einer Sekunde erkannt? (II) Z. Sinnesphysiol. **65** : 248, 1934.

Grossman, F. S.: Studies on the physiology of trace conditioned salivary reflexes. Dissert. St. Petersburg, 1909. (In Russian)

Guilford, J. P.: Ocular movements and the perception of time. J. exp. Psychol. **12** : 259, 1929.

Gulliksen, H.: The influence of occupation upon the perception of time. J. exp. Psychol. **10** : 52, 1927.

Gurevich, B. Kh.: On the conditions of appearance and retention of a dominant respiratory rhythm in the electrocorticogram of the normal rabbit. Fiziol. Zh. **34** : 339, 1948. (In Russian)

Gurevich, B. Kh.: On the correlation of the cortical alpha-rhythm with the respiratory rhythm in the normal rabbit. Fiziol. Zh. **35** : 373, 1949. (In Russian)

Hagbarth, K.-E., Höjeberg, S.: Evidence for subcortical regulation of the afferent discharge to the somatic sensory cortex in man. Nature (Lond.) **179** : 526, 1957.

Halstead, W. C., Knox, G. W., Walker, A. E.: Modification of cortical activity by means of intermittent photic stimulation in the monkey. J. Neurophysiol. **5** : 349, 1942a.

Halstead, W.C., Knox, G. W., Woolf, J. I., Walker, A. E.: Effects of intensity and wave length on driving cortical activity in monkeys. J. Neurophysiol. **5** : 483, 1942b.

Harton, J. J.: The influence of the difficulty of activity on the estimation of time. J. exp. Psychol. **23** : 270, 428, 1938.

Harton, J. J.: An investigation of the influence of success and failure on the estimation of time. J. gen. Psychol. **21** : 51, 1939.

Harton, J. J.: Time estimation in relation to goal organization and difficulty of tasks. J. gen. Psychol. **27** : 63, 1942.

Hawickhorst, L.: Mit welcher Sicherheit wird der Zeitwert einer Sekunde erkannt? (I) Z. Sinnesphysiol. **65** : 58, 1934.

Hernández-Péon, R., Jouvet, M., Scherrer, H.: Auditory potentials at cochlear nucleus during acoustic habituation. Acta neurol. Latinoamer. **3** : 144, 1957.

Hernández-Péon, R., Scherrer, H.: Inhibitory influence of brain stem reticular formation upon synaptic transmission in trigeminal nucleus. Feder. Proc. **14** : 521, 1955.

Hernández-Péon, R., Scherrer, H., Jouvet, M.: Modification of electric activity in cochlear nucleus during "attention" of unanesthetized cats. Science **123** : 331, 1956.

Hernández-Péon, R., Scherrer, H., Velasco, M.: Central influences on afferent conduction in the somatic and visual pathways. Acta neurol. Latinoamer. **2** : 8, 1956.

Hill, C. J., Calvin, J. S.: The joint extinction of two simple excitatory tendencies. J. comp. Psychol. **27** : 215, 1939.

Hill, D., Parr, G., ed.: Electroencephalography. A symposium on its various aspects. London 1950.

Hoagland, H.: Pacemakers in relation to aspects of behavior. New York 1935.

Hoagland, H.: Pacemakers of human brain waves in normals and in general paretics. Amer. J. Physiol. **116** : 604, 1936.

Hoffmann, K.: Experimentelle Aenderung des Richtungsfindens beim Star durch Beeinflussung der „inneren Uhr." Naturwiss. **40** : 608, 1953.

Holubář, J.: Stimulator for study of neural action potentials. Biol. Listy **30** : 26, 1949. (In Czech)

Holubář, J.: Heart electric potentials in the volume conductor. Physiol. Bohemoslov. **5** (Suppl.) : 22, 1956.

Holubář, J.: Recent findings on the physiology of the reticular formation. Čs. Fysiol. **6** : 262, 1957. (In Czech)

Holubář, J.: Electroencephalographic manifestations of the galvanic skin reflex in man. Čs. Fysiol. **7** : 179, 1958a. (In Czech)

Holubář, J.: An electroencephalographic contribution to the function of the reticular formation in man. Čs. Fysiol. **7** : 470, 1958b. (In Czech)

Holubář, J.: EEG manifestations of the unconditioned and conditioned skin galvanic response. EEG clin. Neurophysiol. **11** : 177, 1959a.

Holubář, J.: A simple resistance record as a pneumograph. Čs. Fysiol. **8** : 119, 1959b. (In Czech)

Holubář, J.: Imposed rhythms in the EEG and time sense. Čs. Fysiol. **8** : 197, 1959c. (In Czech)

Holubář, J.: Imposed rhythms in the EEG and the respiratory rhythm. Čs. Fysiol. **8** : 408, 1959d. (In Czech)

Holubář, J.: The time sense and photic driving. EEG clin. Neurophysiol. **12** : 533, 1960.

Holubář, J.: Electroencephalographic manifestations of the skin galvanic reflex in man. Physiol. Bohemoslov. **9** : 472, 1960.

Holubář, J.: Skin galvanic responses and accompanying EEG changes as conditioned reflex to time. Physiol. Bohemoslov. **9** : 477, 1960.

Holubář, J.: The time sense and photic driving in man. Physiol. Bohemoslov **9** : 482, 1960.

Holubář, J.: Navigation of birds—the solved problem of the "sixth sense". Vesmír **39** : 13, 1960; Postovní Holubářství **5** : 22, 1960. (In Czech)

Holubář, J.: On time sense—the so-called biological clock. Vestmír **39** : 201, 1960. (In Czech)

Holubář, J., Kohlik, E.: On the question of induced rhythms in the retina and brain. Čas. Lek. čes. **89** : 974, 1950. (In Czech)

Holubář, J., Machek, J.: Time sense during epileptic patterns in the EEG. Čs. Fysiol. **9** : 422, 1960. (In Czech)

Homack, W.: Ueber das subjektive Abgrenzen von Intervallen. I. Inaug.-Dissert. Jena 1935.

Hormia, A.: On the sensation of duration. Ann. Acad. Sci. Fenn., Ser. A, V, No. 58. Helsinki 1956.

Hull, C. L.: Principles of behavior. New York, London 1943.

Hülser, C.: Zeitauffassung und Zeitschätzung verschieden ausgefüllter Intervalle unter besonderer Berücksichtigung der Aufmerksamkeitsablenkung. Arch. ges. Psychol. **49** : 363, 1924.

Israeli, N.: The psychopathology of time. Psychol. Rev. **39** : 486, 1932.

Jaensch, E. R., Kretz, A.: Experimentell-struktur psychologische Untersuchungen über die Auffassung der Zeit unter Berücksichtigung der Personaltypen. Z. Psychol. **126** : 312, 1932.

Janet, P.: L'évolution de la mémoire et de la notion du temps. Paris 1928.

Jasper, H. H.: Electrical signs of cortical activity. Psychol. Bull. **34** : 411, 1937.

Jasper, H. H., Shagass, C.: Conditioning the occipital alpha rhythm in man J. exp. Psychol. **28** : 373, 1941a.

Jasper, H. H., Shagass, C.: Conscious time judgements related to conditioned time intervals and voluntary control of alpha rhythm. J. exp. Psychol. **28** : 503, 1941b.

Kahnt, O.: Ueber den Gang des Schätzungsfehlers bei der Vergleichung von Zeitstrecken. Psychol. Stud. **9** : 279, 1914.

Kanaev, I. I.: Studies on the physiology of time estimation in children. Fiziol. Zh. **42** : 341, 1956. (In Russian)

Kayser, C.: Le sommeil. J. Physiol. (Paris) **41** : 1 A, 1949.

Köhler, W.: Zur Theorie des Sukzessivvergleichs und der Zeitfehler. Psychol. Forsch. **4** : 115, 1923.

Koehnlein, H.: Über das absolute Zeitgedächtnis. Z. Sinnesphysiol. **65** : 35, 1934.

von Kries, J.: Über die Bedeutung des Aufmerksamkeitsursprunges für den Zeitsinn. Z. Nervenheilk. **47–48** : 352, 1913.

von Kries, J.: Allgemeine Sinnesphysiologie. Leipzig 1923.

Krzhishkovskiĭ, K. N.: On the physiology of conditioned

inhibition. Trudy Ob. Rus. Vrach. St. Petersburg, 1908. (In Russian)

Kvasnitskiĭ, A. V., Koniukhova, V. A.: Application of the studies of I. P. Pavlov to animal husbandry. Kiev, 1954 (cited by Dmitriev and Kochigina 1955). (In Russian)

Larsson, L. E.: The relation between the startle reaction and the nonspecific EEG response to sudden stimuli, with a discussion on the mechanisms of arousal. EEG clin. Neurophysiol. **8** : 631, 1956.

Leonow, W. A.: Über die Bildung von bedingten Spurenreflexen bei Kindern. Pflügers Arch. ges. Physiol. **214** : 305, 1926.

Libet, B., Gerard, R. W.: Control of the potential rhythm of the isolated frog brain. J. Neurophysiol. **2** : 153, 1939.

Livingston, R. B.: Central control of afferent activity. In: Jasper, H. H. et al., ed.: Reticular formation of the brain Boston 1958. (p. 177).

Loomis, A. L., Harvey, E. N., Hobart, G. A.: Potential rhythms of the cerebral cortex during sleep. Science **81** : 597, 1935a.

Loomis, A. L. et al.: Further observations on potential rhythms of cerebral cortex during sleep. Science **82** : 198, 1935b.

Loomis, A. L. et al.: Electric potentials of the human brain. J. exp. Psychol. **19** : 249, 1936.

Loomis, A. L. et al.: Cerebral states during sleep as studied by brain potentials. J. exp. Psychol. **21** : 127, 1937.

Loomis, A. L. et al.: Distribution of disturbance patterns in the human EEG with special reference to sleep. J. Neurophysiol. **1** : 413, 1938.

Maiorov, F. P.: The more complex aspects of the physiology of higher nervous activity. Trudy fiziol. Lab. im. I. P. Pavlova **5** : 255, 1933. (In Russian)

McClelland, D. C.: Factors influencing the time error in judgments of visual extent. J. exp. Psychol. **33** : 81, 1943.

Matthews, G. V. T.: Bird navigation. Cambridge 1955.

Mehner, M.: Zur Lehre vom Zeitsinn. Wundts philos. Stud. **2** : 546, 1885.

Meumann, E: Beiträge zur Psychologie des Zeitsinns. Wundts philos. Stud. **8** : 431, 1893.

Meumann, E.: Beiträge zur Psychologie des Zeitsinns. Wundts philos. Stud. **9** : 264, 1894.

Meumann, E.: Beiträge zur Psychologie des Zeitbewusstseins. Wundts philos. Stud. **12** : 125, 1896.

Münsterberg, H.: Beiträge zur experimentellen Psychologie. Heft 1, 2. Freiburg 1889.

Nikiforovskiĭ, P. M.: On the physiology of time. Rus. fiziol. Zh. **12** : 483, 1929. (In Russian)

Nikiforovskiĭ, P. M.: Some aspects of the study of higher nervous activity. Zh. Vyssh. Nerv. Deiat. **1** : 827, 1951. (In Russian)

Omwake, K. T., Loranz, M.: Study of ability to wake at a specified time. J. appl. Psychol. **17** : 468, 1933.

Oswald, I.: A proposed origin of the non-specific EEG response. EEG clin. Neurophysiol. **11** : 341, 1959.

Pampiglione, M. C.: The phenomenon of adaptation in human EEG (a study of K-complexes) Rev. neurol. (Paris) **87** : 197, 1952.

Pauli, R.: Psychologisches Praktikum. Jena 1950.

Pavlov, I. P.: Complete collected works, III. 2nd. ed. Moscow 1951–52. (In Russian)

Philip, B. R.: Time errors in the discrimination of color mass by the ranking method. J. exp. Psychol. **27** : 285, 1940.

Philip, B. R.: The effect of interpolated and extrapolated stimuli on the time order error in the comparison of temporal intervals. J. gen. Psychol. **36** : 173, 1947.

Pimenov, P. P.: A special group of conditioned reflexes. Dissert. St. Petersburg, 1907. (In Russian)

Quasebarth, K.: Zeitschätzung und Zeitauffassung optisch und akustisch ausgefüllter Intervalle. Arch. ges. Psychol. **49** : 379, 1924.

Rodnick, E. H.: Does the interval of delay of conditioned responses possess inhibitory properties? J. exp. Psychol. **20** : 507, 1937.

Roth, B.: Narcolepsy and hypersomnia from the standpoint of the physiology of sleep. Prague, 1957. (In Czech)

Roth, M., Shaw, J., Green, J.: The form, voltage distribution and physiological significance of the K-complex. EEG clin. Neurophysiol. **8** : 385, 1956.

Schaefer, V. G., Gilliland, A. R.: The relation of time estimation to certain physiological changes. J. exp. Psychol. **23** : 545, 1938.

Scherrer, H., Hernández-Péon, R.: Inhibitory influence of reticular formation upon synaptic transmission in gracilis nucleus. Feder. Proc. **14** : 132, 1955.

Schulz, B.: Über fortlaufende Zeitschätzungen. Psychol. Arb. **9** : 120, 1927.

Schumann, F.: Über die Schätzung kleiner Zeitgrössen. Z. Psychol. **4** : 1, 1893.

Schütz, E., Müller, H. W., Schönenberg, H.: Über die Entwicklung zentralnervöser Rhythmen im Elektroencephalogramm des Kindes. Z. ges. exp. Med. **117** : 157, 1951.

Simonson, E., Brožek, J.: Flicker fusion frequency. Background and applications. Physiol. Rev. 32 : 349, 1952.

von Skramlik, E.: Die Angleichung der subjektiven Zeitauffassung an astronomische Vorgänge. Die physiologische Uhr. Naturwiss. 22 : 98, 1934a.

von Skramlik, E.: Objektive und subjektive Zeiteinheit. Klin. Wschr. 13 : 433, 1934b.

von Skramlik, E.: Die physiologische Uhr. Münch. med. Wschr. 82 : 485, 1935.

von Skramlik, E.: Die Erlebniszeit, ihre Festsetzung und Einteilung. Z. physik. chem. Unterricht 52 : No 6, 1939.

Spiegel, E. A., Wycis, H. T., Orchinik, C. W., Freed, H.: The thalamus and temporal orientation. Science 121 : 771, 1955.

Sterzinger, O.: Chemopsychologische Untersuchungen über den Zeitsinn. Z. Psychol. 134 : 100, 1935.

Sterzinger, O.: Neue chemopsychologische Untersuchungen über den menschlichen Zeitsinn (das Problem der 5-Minuten-Zeitstrecke). Z. Psychol. 143 : 391, 1938.

Straus, E.: Das Zeiterlebnis in der endogenen Depression und in der psychopathischen Verstimmung. Mschr. Psychiat. Neurol. 68 : 640, 1928.

Stott, L. H.: Time-order errors in the discrimination of short tonal durations. J. exp. Psychol. 18 : 741, 1935.

Stukova, M. M.: Further studies on the physiology of time as a conditional stimulus of the salivary glands. Dissert. St. Petersburg, 1914. (In Russian)

Tarchanoff, J.: Über die galvanischen Erscheinungen in der Haut des Menschen bei Reizung der Sinnesorgane und bei verschiedenen Formen der psychischen Tätigkeit. Pflügers Arch. ges. Physiol. 46 : 46, 1890.

Toman, J.: Flicker potentials and the alpha rhythm in man. J. Neurophysiol. 4 : 51, 1941.

Tresselt, M. E.: Time errors in successive comparison of simple visual objects. Amer. J. Psychol. 57 : 555, 1944a.

Tresselt, M. E.: The time errors in visual extents and areas. J. Psychol. 17 : 21, 1944b.

Vatsuro, E. G.: The reflex to time in the scheme of conditional stimuli. Trudy fiziol. Lab. im. I. P. Pavlova 13 : 5, 1948. (In Russian)

Vasilenko, F. D.: On the question of the conditioned reflex to time. Trudy fiziol. Lab. im. I. P. Pavlova 4 : 310, 1932. (In Russian)

Veraguth, O.: Das psychogalvanische Reflexphänomen. Berlin 1909.

Vierordt, K.: Der Zeitsinn nach Versuchen. Tübingen 1868.
Vondráček, V.: Pharmacology of the mind. Prague 1935. (In Czech)
Vondráček, V.: Perception. Prague, 1949. (In Czech)
Walker, A. E., Woolf, J. I., Halstead, W. C., Case, T. J.: Mechanism of temporal fusion effect of photic stimulation on electrical activity of visual structures. J. Neurophysiol. 6 : 213, 1943.
Walker, A. E. et al.: Photic driving. Arch. Neurol. Psychiat. (Chicago) 52 : 117, 1944b.
Walsh, E. G.: Autonomy of alpha rhythm generators studied by multiple channel cross-correlation. EEG clin. Neurophysiol. 10 : 121, 1958.
Walter, V. J., Walter, W. G.: Central effects of rhythmic sensory stimulation. EEG clin. Neurophysiol. 1 : 57, 1949.
Wang, G. H., Brown, V. W.: Changes in galvanic skin reflex after acute spinal transection in normal and decerebrate cats. J. Neurophysiol. 19 : 446, 1956a.
Wang, G. H., Brown, V. W.: Suprasegmental inhibition of an autonomic reflex. J. Neurophysiol. 19 : 564, 1956b.
Wang, G. H., Stein, P., Brown, V. W.: Brainstem reticular system and galvanic skin reflex in acute decerebrate cats. J. Neurophysiol. 19 : 350, 1956.
Webb, H. M., Brown, F. A. Jr.: Timing long-cycle physiological rhythms. Physiol. Rev. 39 : 127, 1959.
Wirth, W.: Die unmittelbare Teilung einer gegebenen Zeitstrecke. Amer. J. Psychol. 50 : 79, 1937.
Woodrow, H.: A quantitative study of rhythm. Arch. Psychol. (New York) 18 : 1, 1909.
Woodrow, H.: The reproduction of temporal intervals. J. exp. Psychol. 13 : 473, 1930.
Woodrow, H.: Individual differences in the reproduction of temporal intervals. Amer. J. Psychol. 45 : 271, 1933.
Woodrow, H.: Time perception. In: Stevens S. S., ed.: Handbook of experimental psychology. New York, London, 1951. (p. 1224).
Wundt, W.: Grundzüge der physiologischen Psychologie. Leipzig 1903. (Bd. 3).
Yoshii, N., Pruvot, P., Gastaut, H.: A propos d'une activité rythmique transitoirement enregistrée dans la formation réticulée mésencéphalique et susceptible de représenter l'expression électroencephalographique de la trace mnémonique. C. R. Acad. Sci. (Paris) 242 : 1361, 1956.
Yoshii, N. et al.: Electrographic activity of the mesencephalic

reticular formation during conditioning in the cat. EEG clin. Neurophysiol. **9** : 595, 1957.

Youtz, R. E. P.: The weakening of one Thorndikian response following the extinction of another. J. exp. Psychol. **24** : 294, 1939.

Zavadskiĭ, I. V.: Studies on the problem of inhibition and extinction of conditioned reflexes. Dissert. St. Petersburg, 1908. (In Russian)

Zeleniĭ, G. P.: Studies on the question of the response of the dog to auditory stimuli. Dissert. St. Petersburg, 1907. (In Russian)

Zeleniĭ, G. P.: On rhythmical muscular movements. Russk. fiziol. Zh. **6** : 155, 1923. (In Russian)

SUPPLEMENTARY BIBLIOGRAPHY

Anan'ev, B. G., Lomov, B. F., ed.: Perception of space and time. Leningrad, Leningrad State University Press, 1961. (In Russian)

Andersen, P., Andersson, S.: Physiological basis of the alpha rhythm (Neuroscience Series #1). New York, Appleton-Century Crofts, 1968.

Anliker, J.: Variations in alpha voltage of the electroencephalogram and time perception. Science **140** : 1307, 1963.

Aschoff, J., ed.: Circadian clocks: Proceedings of the Feldafing Summer School, 1964. Amsterdam, North-Holland Publishing Co., 1965.

Baker, C. H.: On temporal extrapolation. Canad. J. Psychol. **16** : 37, 1962.

Barlow, J. S.: Autocorrelation and crosscorrelation techniques in EEG analysis. In: Computer techniques in EEG analysis, Electroenceph. clin. Neurophysiol. Suppl. **20** : 31, 1961.

Barlow, J. S.: The relationship among "photic driving" responses to single flashes, and the resting EEG. Quart. Prog. Rept. No. 64, Res. Lab. Electron., M.I.T.: 274, 1962.

Barlow, J. S.: Evoked responses in relation to visual perception and oculomotor reaction times in man. Ann. N. Y. Acad. Sci. **112**, Art. 1 : 432, 1964.

Barlow, J. S.: Inertial navigation as a basis for animal navigation. J. Theoret. Biol. **6** : 76, 1964.

Barlow, J. S.: Inertial navigation and animal navigation. New Eng. J. Med. **273** : 1090, 1965.

Barlow, J. S.: Inertial navigation in relation to animal navigation. J. Inst. Navig. **19** : 302, 1966.

Barlow, J. S.: Observations on the electrophysiology of timing in the brain. Electroenceph. clin. Neurophysiol. 1969 (In press).

Beer, B., Trumbule, G.: Timing behavior as a function of amount of reinforcement. Psychon. Sci. **2** : 71, 1965.

Behar, I.: A method for scaling in infrahuman species: time perception in monkeys. Percept. Motor Skills **16** : 275, 1963.

Bell, C. R.: Control of time estimations by a chemical clock. Nature **210** : 1189, 1966.

Bellrose, F. C.: Radar in orientation research. Proc. XIV Int. Orn. Cong. Oxford, 1967, 281.

Bokander, I.: Time estimation as an indicator of attention-arousal when perceiving complex and meaningful stimulus material. Percept. Motor Skills **21** : 323, 1965.

Braitenberg, V.: Is the cerebellar cortex a biological clock in the millisecond range? In: Progress in brain research, Vol. 25, The cerebellum. Amsterdam, Elsevier Co., 1967, 334.

Braitenberg, V., Onesto, N.: The cerebellar cortex as a timing organ; discussion of an hypothesis. Atti del 1° Congresso Internazionale di Medicina Cibernetica. Naples, Giannini, 1962, 3.

Brazier, M. A. B., ed.: Computer techniques in EEG analysis. Electroenceph. clin. Neurophysiol., Suppl. 20, 1961.

Bünning, E.: The physiological clock. New York, Academic Press, 1964.

Chatterjea, R. G., Rakshit, P.: Estimation of temporal interval. Percept. Motor Skills **22** : 176, 1966.

Cohen, J.: Psychological time. Sci. Amer. **211** : Nov., 116, 1964.

Cohen, J.: Psychological time in health and disease. Springfield, Ill., Charles C Thomas, 1967.

Cold Spring Harbor Symposia on Quantitative Biology, Vol. 25 (Biological clocks). Cold Spring Harbor, L. I., New York, The Biological Laboratory, 1961.

Creutzfeldt, O. D., Watanabe, S., Lux, H. D.: Relations between EEG phenomena and potentials of single cortical cells. II. Spontaneous and convulsoid activity. Electroenceph. clin. Neurophysiol. **20** : 19, 1966.

Danziger, K., duPreez, P. D.: Reliability of time estimations by the method of reproduction. Percept. Motor Skills **16** : 879, 1963.

Davis, R.: Choice reaction times and the theory of intermittency in human performance. Quart. J. Exper. Psychol. **14**, Part 3: 157, 1963.

Davis, R.: Time uncertainty and the estimation of time-intervals. Nature **195** : 311, 1962.

DeNike, L. D.: The temporal relationship between awareness and performance in verbal conditioning. J. exp. Psychol. **68** : 521, 1964.

Denner, B., Wapner, S., McFarland, J. H., Werner, H.: Rhythmic activity and the perception of time. Amer. J. Psychol. **76** : 287, 1963.

Dimond, S. J.: The structural basis of timing. Psychol. Bull. **62** : 348, 1964.

Dinnerstein, A. J., Zlotogura, P.: Intermodal perception of temporal order and motor skills: effects of age. Percept. Motor Skills **26** : 987, 1968.

Dmitriev, A. S.: On physiological mechanisms of time perception. Usp. Sovr. Biol. **52** : 274, 1961. (In Russian)

Dmitriev, A. S.: The physiological basis of human time perception. Usp. Sovr. Biol. **57** : 245, 1964. (In Russian)

Dmitriev, A. S.: Analysis of the process of formation of conditioned reflexes to time. Zh. vyssh. nerv. Deiat. **14** : 618, 1964. (In Russian)

Dmitriev, A. S., Semenov, V. N.: Age peculiarities of conditioned reflexes to time. Zh. vyssh. nerv. Deiat. **11** : 723, 1961. (In Russian)

Doering, D. G.: Accuracy and consistency of time-estimation by four methods of reproduction. Amer. J. Psychol. **74** : 27, 1961.

Dorst, J.: The migrations of birds. Boston, Houghton Mifflin Co., 1962.

duPreez, P.: Field dependence and accuracy of comparison of time intervals. Percept. Motor Skills **24** : 467, 1967.

duPreez, P.: Reproduction of time intervals after short periods of delay. J. gen. Psychol. **76** : 59, 1967.

Efron, R.: Temporal perception, aphasia and déjà vu. Brain **86** : 403, 1963.

Efron, R.: The effect of stimulus intensity on the perception of simultaneity in right- and left-handed cases. Brain **86** : 285, 1963.

Efron, R.: The effect of handedness on the perception of simultaneity and temporal order. Brain **86** : 261, 1963.

Efron, R.: The minimum duration of a perception. Neuropsychol. 1969 (in press).

Elkin, D. G.: Perception of time. Moscow, RSFR, Acad. Pedag. Sci., 1962. (In Russian)

Emlen, S. T.: Migratory orientation in the indigo bunting, *Passerina cyanea.* Part II: Mechanism of celestial orientation. Auk. 84 : 463, 1967.

Emley, G. S., Schuster, C. R., Lucchesi, B. R.: Trends observed in time estimation of three stimulus intervals within and across sessions. Percept. Motor Skills 26 : 391, 1968.

Fischer, R., consulting ed.: Interdisciplinary perspectives of time. Ann. New York Acad. Sci. 138, Art. 2 : 367, 1966.

Fitzgerald, H. E., Lintz, L. M., Brackbill, Y., Adams, G.: Time perception and conditioning and autonomic response in human infants. Percept. Motor Skills 24 : 479, 1967.

Fraisse, P.: Influence de la fréquence sur l'estimation du temps. Année Psychol. 61 : 325, 1961.

Fraisse, P.: The psychology of time. New York, Harper and Row, 1963.

Fraser, J. T., ed.: The voices of time. New York, George Braziller, 1966.

Friel, C. S., Lhamon, W. T.: Gestalt study of time estimation. Percept. Motor Skills 21 : 603, 1965.

Goldstone, S.: Variability of temporal judgment: intersensory comparisons and sex differences. Percept. Motor Skills 26 : 211, 1968.

Goldstone, S., Goldfarb, J. L.: Judgment of filled and unfilled durations: intersensory factors. Percept. Motor Skills 17 : 763, 1963.

Gooddy, W.: Disorders of orientation in space-time. Brit. J. Psychiat. 112, No. 488 : 661, 1966.

Grant, R. A., McFarling, C., Gormezano, I.: Temporal conditioning and the effect of interpolated UCS presentation in eyelid conditioning. J. gen. Psychol. 63 : 249, 1960.

Grunbaum, A.: Philosophical problems of space and time. New York, Alfred A. Knopf, 1963.

Hake, H. H.: Temporal experience. Ann. Rev. Psychol. 13 : 160, 1962.

Hale, D. J.: Sequential analysis of effects of time uncertainty on choice reaction time. Percept. Motor Skills 25 : 285, 1967.

Hermann, H. T., Quarton, G. C.: Changes in alpha frequency with change in thyroid hormone level. Electroenceph. clin. Neurophysiol. 16 : 515, 1964.

Hirsh, I. J.: Auditory perception of temporal order. J. Acoust. Soc. Amer. 31 : 759, 1959.

Hirsh, I. J., Sherrick, C. E., Jr.: Perceived order in different sense modalities. J. exp. Psychol. **62** : 423, 1961.

Ho, P.-Y., Ma, M.-C.: The problem of time perception in the discrimination of flash light frequencies. Acta Psychol. Sinica No. 4 : 324, 1962. (In Chinese, with English abstract.)

Hoagland, H.: Some biochemical considerations of time. In: The voices of time (J. T. Fraser, ed.), New York, George Braziller, 1967.

Holubář, J.: Zeitsinn und aufgezwungene EEG-Rhythmen. In: Jenenser EEG-Symposion, Berlin, VEB Verlag Volk und Gesundheit, 1963, 103.

Holubář, J., Machek, J.: Time sense and epileptic EEG activity. Epilepsia **3** : 323, 1962.

Husserl, E.: The phenomenology of internal time consciousness (J. S. Churchill, transl.). Bloomington, Indiana, Indiana University Press, 1964.

Jerison, H. J., Pickett, R. M., Stenson, H. H.: The elicited observing rate and decision processes in vigilance. Hum. Factors **7** : 107, 1965.

Johannsen, D. E.: Perception of time. Ann. Rev. Psychol. **18** : 26, 1967.

John, E. R.: Electrophysiological studies of conditioning. In: The neurosciences (eds. G. C. Quarton, T. Melnechuk, and F. O. Schmitt). New York, The Rockefeller University Press, 1967, 690.

Kahn, P.: Time orientation and perceptual and cognitive organization. Percept. Motor Skills **23** : 1059, 1966.

Kawabata, N.: Scalp responses to photic stimulation by time sequence patterns. Electroenceph. clin. Neurophysiol. **25** : 449, 1968.

Kelm, H.: Consistency of successive time estimates during positive feed-back. Percept. Motor Skills **15** : 216, 1962.

Kenna, J. C., Sedman, G.: The subjective experience of time during lysergic acid diethylamid (LSD-25) intoxication. Psychopharmacologia (Berlin) **5** : 280, 1964.

Khametov, B. G.: On the problem of the transformation of conditioned reflexes to time. Zh. vyssh. nerv. Deiat. **11** : 1106, 1961. (In Russian)

Kleber, R. J., Lhamon, W. T., Goldstone, S.: Hyperthermia, hyperthyroidism and time judgments. J. comp. physiol. Psychol. **56** : 362, 1963.

Kramer, G.: Long-distance orientation. In: Biology and comparative physiology of birds (A. J. Marshall ed.), Vol. 2. New York, Academic Press, 1961.

Lane, H.: Temporal and intensive properties of human vocal responding under a schedule of reinforcement. J. exp. anal. Behav. **3** : 183, 1960.

Lansing, R. W., Trunnel, J. B.: Electroencephalographic changes accompanying thyroid deficiency in man. J. clin. Endocr. Metab. **23** : 470, 1963.

Laties, V. G., Weiss, B., Clark, R. L., Reynolds, M. D.: Overt "mediating" behavior during temporally spaced responding. J. exp. anal. Behav. **8** : 107, 1965.

Lhamon, W. T., Edelberg, R., Goldstone, S.: A comparison of tactile and auditory time judgment. Percept. Motor Skills **14** : 366, 1962.

Lichtenstein, M., White, C. T.: Synchronous tapping and the concept of quantized psychological units of duration. Percept. Motor Skills **18** : 217, 1964.

Lockart, R. A.: Temporal conditioning of GSR. J. exp. Psychol. **71** : 438, 1966.

Matsuda, M., Matsuda, F.: Time estimation by reproduction in schizophrenics. Percept. Motor Skills **25** : 898, 1967.

Matthews, G. V. T.: Bird navigation, 2nd ed. Cambridge, Cambridge University Press, 1968.

McGrath, J. J., O'Hanlon, J. F., Jr.: Temporal orientation and vigilance performance. Acta Psychol. (Amst.) **27** : 410, 1967.

McGrath, J. J., O'Hanlon, J. F., Jr.: A method of measuring the rate of subjective time. Percept. Motor Skills **24** : 1235, 1967.

McGrath, J. J., O'Hanlon, J. F., Jr.: Relationships among chronological age, intelligence, and rate of subjective time. Percept. Motor Skills **26** : 1083, 1968.

Meurice, E., Weiner, H.: Activité électroencéphalographique et électrodermographique parallèle à un comportement humain de reproduction temporelle. Rev. Neurol. **113** : 369, 1965.

Migler, B., Brady, J. V.: Timing behavior and conditioned fear. J. exp. anal. Behav. **7** : 247, 1964.

Millenson, J. R., Hurwitz, H. M.: Some temporal and sequential properties of behavior during conditioning and extinction. J. exp. anal. Behav. **4** : 97, 1961.

Miller, A. R., Frauchiger, R. A., Kiker, V. L.: Temporal experience as a function of sensory stimulation and motor activity. Percept. Motor Skills **25** : 997, 1967.

Morrell, F.: Electrical signs of sensory coding. In: The neurosciences (ed. G. C. Quarton, T. Melnechuk, and F. O. Schmitt), New York, The Rockefeller University Press, 1967, 452.

Nelson, T. M., Bartley, S. H., Jordan, J. F.: Experimental

evidence for the involvement of a neurophysiological mechanism in the discrimination of duration. J. Psychol. **55** : 371, 1963.

Nisbet, I. C. T., Drury, W. H., Jr.: Orientation of spring migrants studied by radar. Bird-Banding **38** : 173, 1967.

Ochberg, F. M., Pollack, I. W., Meyer, E.: Reproductive and estimative methods of time judgment. Percept. Motor Skills **20** : 653, 1965.

Pfaff, D.: Effects of temperature and time of day on time judgments. J. exp. Psychol. **76** : 419, 1968.

Plutchik, R., Schwartz, A. K.: Critical analysis of the problem of time-error. Percept. Motor Skills **27** : 79, 1968.

Priestley, J. B.: Man and time. London, Aldus Books, 1964.

Reynolds, G. S., Catania, A. C.: Temporal discrimination in pigeons. Science **135** : 314, 1962.

Reynolds, H. N., Jr.: Temporal estimation in the perception of occluded motion. Percept. Motor Skills **26** : 407, 1968.

Richards, W.: Time estimates measured by reproduction. Percept. Motor Skills **18** : 929, 1964.

Richter, C. P.: Biological clocks in medicine and psychiatry. Springfield, Ill., Charles C Thomas, 1965.

Richter, C. P.: Psychopathology of periodic behavior in animals and man. In: Comparative psychopathology, New York, Grune and Stratton, 1967, 205.

Robbins, M. C., Kilbride, P. L., Bukenya, J. M.: Time estimation and acculturation among the Baganda. Percept. Motor Skills **26** : 1010, 1968.

Rosner, B. S.: Temporal interaction between electrocutaneous stimuli. J. exp. Psychol. **67** : 191, 1964.

Rusinov, V. S.: Electroencephalographic studies in conditional reflex formation in man. In: The central nervous system and behavior (Trans. Second Conf.) ed. M. A. B. Brazier. New York, Josiah Macy, Jr., Foundation, 1959, 249.

Rutschmann, J., Link, R.: Perception of temporal order of stimuli differing in sense mode and simple reaction time. Percept. Motor Skills **18** : 345, 1964.

Schmidt-Koenig, K.: Current problems in bird orientation. In: Advances in the study of behavior, Vol. 1 (ed. D. S. Lehrman, R. A. Hinds, and E. Shaw). New York, Academic Press, 1965, 217.

Siegman, A. W.: Intercorrelation of some measures of time estimation. Percept. Motor Skills **14** : 381, 1962.

Smart, J. J. C.: Problems of space and time. New York, Macmillan Co., 1964.

Strumwasser, F.: Neurophysiological aspects of rhythms. In: The neurosciences (ed. G. C. Quarton, T. Melnechuk, and F. O. Schmitt). New York, The Rockefeller University Press, 1967, 516.

Surwillo, W. W.: Age and the perception of short intervals of time. J. Geront. **19** : 322, 1964.

Surwillo, W. W.: Time perception and the "internal clock": some observations on the role of the EEG. Brain Res. **2** : 390, 1966.

Talland, G. A.: Deranged memory. New York and London, Academic Press, 1965.

Tanaka, Y.: Time order error. Ann. Rev. Psychol. **17** : 250, 1966.

Treisman, M.: Temporal discrimination and the indifference interval, implications for a model of the "internal clock". Psychol. Monogr. **77** No. 576, 1963.

Verhave, T.: Avoidance responding as a function of simultaneous and equal changes to two temporal parameters. J. exp. anal. Behav. **2** : 185, 1959.

Vetter, K., Horvath, S. M.: Effect of audiometric parameters on K-complex of electroencephalogram. Psychiat. Neurol. (Basel) **144**, 103, 1962.

Victor, M., Angevine, J. B., Mancall, E. L., Fisher, C. M.: Memory loss with lesions of hippocampal formation. Arch. Neurol. **5** : 244, 1961.

Wallraff, H. G.: The present status of our knowledge about pigeon homing. In: Proceedings of the XIV International Ornithological Congress (ed. D. W. Snow). Oxford and Edinburgh, Blackwell Scientific Publications, 1967, 331.

Wallraff, H. G.: Direction training of birds under a planetarium sky. Naturwiss. **55** : 235, 1968.

Watanabe, S.: The cybernetical view of time. Progr. Biocybernet. **3** : 152, 1966.

Weber, D. S.: A time perception task. Percept. Motor Skills **21** : 863, 1965.

Werboff, J.: Time judgment as a function of electroencephalographic activity. Exp. Neurol. **6** : 152, 1962.

White, C. T.: Temporal numerosity and the psychological unit of duration. Psychol. Monogr. **77** No. 575, 1963.

White, C. T., Lichtenstein, M.: Some aspects of temporal discrimination. Percept. Motor Skills **17** : 471, 1963.

Whyman, A. D., Moos, R. H.: Time perception and anxiety. Percept. Motor Skills **24** : 567, 1967.

Wiener, N.: Application to the study of brain waves, random time, and coupled oscillators (Lecture 8). In: Nonlinear problems in random theory, Cambridge, Mass., and New York, Technology Press of the Massachusetts Institute of Technology and John Wiley & Sons, 1958, 67.

Wiener, N.: Brain waves and self-organizing systems. Ch. 10 In: Cybernetics (2nd ed.). Cambridge, Mass., and New York, The M.I.T. Press and John Wiley & Sons, 1961, 181.

EPILOGUE

The following commemorative tribute to Josef Holubář, M.D., Sc.D. (1917–1967) by Professor Zdenek Servít, Chief of the Institute of Physiology at the Czechoslovak Academy of Sciences, appeared in Československá Fysiologie (Volume 17, 1968, page 163), together with a complete list of the more than seventy publications by Holubář in scientific journals.

In Memoriam

Often it is only after a man has departed from our midst that we fully appreciate the extent of his importance and the magnitude of the void that remains after him. This is particularly so for modest and unassuming individuals, the more so when death is as tragic and unexpected as it was for Josef Holubář.

His career in physiology began at the Institute of Physiology of Charles University in Prague in the first years following the war. After the establishment of the

Czechoslovak Academy of Sciences he worked in the laboratory of Academician V. Laufberger, and for the last eight years in the Institute of Physiology of the Czechoslovak Academy of Sciences. He did not interrupt his own academic work, however. At the Academy, he received the degree of doctor of science, and at Charles University the faculty title of "docent" (associate professor). His first publication did not appear until the year 1948; but then, in less than twenty years, he published almost 70 papers, several textbooks, a section on neurophysiology and sensory physiology in a textbook on physiology for medical students, a monograph, "The Sense of Time," and an extensive series of popular scientific articles.

Although he had a wide range of knowledge covering the whole field of physiology, the area of his own research was neurophysiology. He began with the physiology of the peripheral nerve, the axon. A series of papers were then devoted to conditioned neurovegetative mechanisms, particularly in connection with physical exercise, and the electrophysiology of temporary connections. It was in this way that he arrived at the very original concept of the alpha rhythm as an internal time-measure for estimation of short time intervals. This work was highly regarded by the founder of cybernetics, Norbert Wiener, when the latter visited Prague. Its compilation as a monograph constituted Holubář's doctoral dissertation.

During the years of his work at the Institute of Physiology of the Czechoslovak Academy of Sciences, he devoted himself to the physiology of intrinsic and evoked potentials of the cerebral cortex and to the pathophysiology of the epileptogenic focus.

Holubář deserves great credit in the methodological development of our neurophysiology. Already in the first years of his career he did valuable work in the application of mathematical methods to physiological investigations. He accomplished a great deal in electrophysiology—we can justly consider him as one of the founders of modern Czechoslovak electrophysiology. Nor must his pedagogical contributions be underestimated, not only as a university teacher in lectures and courses, but also—and perhaps even more—as an unselfish advisor on methodology and collaborator of anyone who turned to him. We all have learned from him.

His most recent significant work on methodology had concerned the identification of individual nerve cells in the cerebral cortex by staining with the same intracellular microelectrode used for recording their electrical activity. I believe this will be a significant step forward in the cytophysiology of the neuron. Unfortunately, this work was not published until after Holubář's death.

Holubář died at the peak of his scientific productivity. Had he been less modest and more forward, he would be better known both here and abroad, and his name would be more frequently cited. As it is, there remains less personal fame in his wake, but the same number of achievements and a warmer recollection among those who knew him and who were taught by him.

To his work we will be bound at every step. This, I believe, is the greater and the more enduring remembrance.

Z. Servít

INDEX

Acceleration, biological effects of, 2
Adrian, E. D., 30, 76
  and Matthews, 30
Afferent system, nonspecific, and K wave, 69
Alekseev, M. A., 9
Alpha activity, and time-measuring mechanism, 77
Alpha blocking, temporal conditioning of, v
Alpha frequency, increase of, with elevated temperature, v
Alpha rhythm, v
  blocking of, 56
    See also Alpha blocking
  conditioned blocking of, 9
  and corticothalamic circulation of impulses, 77
  frequency of, and flicker frequency, 74–75
  and interconnection of brain structures, 77
  multiple generators of, 77, 78
  origin of, 76–78
  and thalamo-cortical connections, 78
  and rhythmic oscillations of neurons, 77
  and sense of time, 76, 78
  and time-measuring mechanism, 78
  as a time-measuring rhythm, vi, 75–76, 83
  in water beetle, 77
  See also Alpha activity; Alpha frequency; Brain rhythms; EEG; EEG rhythms
Anesthesia, local, 80
Arvanitaki, A., 77
Asratian, E. A., 69

von Baer, K., 17
Baiandurov, B. I., 8
Bancaud, J., et al., 54
Bartley, S. H., 30
  and Bishop, 30
Bekhterev, V. M., 8
Bekkering, D. H., et al., 78

Benussi, V., 15, 16
Beritov, I. S., 8
Binet, A., and Simon, 19, 76
Biological clock(s)
 autonomous rhythms as, 28
 effect of flicker on, 41
 and hibernation, 24
 metabolic basis of, 19
 pendulum of, 74
Biological rhythm(s)
 autonomous, heart beat as an, 29
 and environmental rhythms, 27
 and pharmacological agents, 24
 range of frequencies of, 27
 and temperature in ants, 24
Biological time, see Time, biological; Sense of time
Birds
 "chronometer" of, 11, 13
 navigation of, see Navigation of birds
 orientation of, and artificial sun, 13
 shift with shift of internal time, 13
 time sense of, 11, 13
Bishop, G. H., see Bartley and Bishop
Bloch, V., see Bancaud et al.
Body temperature, see Temperature, body
Bolotina, O. P., 8
Brain, 74, 76
 rhythmic activity of, 83
 rhythms
  autonomous, 29
  and flicker, 30
  ontogenesis of, 76
  as pacemakers for sense of time, 30
  and sense of time, 74
  temperature dependence of, 76
  as a time-measuring pendulum, 85
  See also Alpha activity; Alpha rhythm; EEG; EEG rhythm(s)
 stem
  and galvanic skin response, 62
  and K wave, 68
 time-measuring (timing)
  mechanism of, v, 75
 "Brain clocks," vi
Brazier, M. A. B., 76
Bremer, F., 77
Bridge circuit (for skin resistance), 35
Brown, F. A., Jr., see Webb and Brown
Brown, V. W., see Wang et al.
Brožek, J., see Simonson and Brožek

Calvin, J. S., see Hill, C. J., and Calvin
Čapek, D., 2
Carrel, A., 19
Case, T. J., see Walker et al.
Central nervous system, and galvanic skin reflex, 62
Cerebral cortex, 81
 and galvanic skin reflex, 62
 and sense of time, 25
 See also Cortex
Chronometer, of birds, 13
Chronometry, physiological, vii
 See also Clocks; Pendulums
Clock(s), 4, 5
 the body as a, 22
 definition of a, 23
 pendulum of, 22
Conditioned inhibition, of galvanic skin reflex, 69
Conditioned reflex(es), 42, 63, 66
 delayed, 69
 reinforcement of, 6
 to time, see Temporally conditioned reflexes
 See also Temporally conditioned reflexes; Trace-conditioned reflexes
Cooper, L. F., and Erickson, 21
Cortex, 67
 isolated, rhythmic activity in, 77

Cortical projection
 of retina, 4
 of somatic sensation, 4
Cortical response, secondary,
 and K wave, 67
Critical fusion frequency, 17
Crosscorrelation, 85
Czechoslovak Academy of
 Sciences, viii, 109, 110

Davis, H., 2
 et al., 47
Davis, P. A., 47
 see also Davis, H., et al.
Davis, R. C., 58, 61
Dempsey, E. W., and Morison, 77
Deriabin, V. S., 7
Diffuse projection system,
 and K waves, 68
Disinhibition, nonspecific, of
 temporally conditioned
 reflexes, 73
Dmitriev, A. S., and Kochigina, 8, 9, 15
Dobrovolskii, F. M., 7
Dreams, and time perception, 82
Drechsler, B., 38
 and Škorpil, 62
Durup, G., and Fessard, 30
Dusser de Barenne, J. G., and
 McCulloch, 77
Dvořák, J., 2

EEG (Electroencephalogram),
 v, 33, 82
 crosscorrelation with flicker, 85
 and flicker, vii
 frequency analysis of, 76, 85
 generalized changes of, and
  galvanic skin reflex, 56
 photic driving (entrainment)
  of, vi
 physiological basis of, vii
 recording of, 31
 rhythms
  and body temperature, 24
  imposed, 75
  and respiratory rhythm, 80
  subcortical propagation of,
   in the rabbit, 80
  in rabbit, and respiratory
   rhythm, 79
  See also Alpha activity;
   Alpha rhythm; Brain
   rhythms
Ehrenwald, H., 25
Einstein, A., 1, 23
 and Infeld, 23
Ejner, M., 20, 21, 22
EKG (electrocardiogram),
 recording of, 31, 37, 38
Electrodes
 EEG, 37
 galvanic skin reflex, 37
 screw, 80
 for stimulation, 34
Electroencephalograph, 36
Electroencephalography, and
 sense of time, ix
 See also EEG
Electrophysiology, vii
Ellson, D. G., 69
Epilepsy, photogenic, 81
Erickson, M. H., see Cooper
 and Erickson
Estel, V., 17
Estimation of time (intervals),
 see Time intervals,
 estimation of
Extinction, secondary, 69

Feokritova, J. P., 6, 7
Ferrari, G. C., 20
Fessard, A., see Durup and
 Fessard
Flicker, vi, 39
 and biological clock(s), 41
 and brain rhythms, 30
 and EEG, vii
 and galvanic skin reflexes, vi
 and heart and respiratory
  rates in rabbits, 32, 80
 method of presentation of,
  31–32, 35

Flicker *(continued)*
 and navigation of birds, 85
 and production and reproduction of a rhythm (tempo), 39–41
 prolongation of intervals of temporally conditioned reflexes by, 78
 and pulse and respiratory rates, 78
 rhythms imposed by, 84
  See also EEG rhythms, imposed
 and sense of time, 75
 and sleep, 85
 and temporally conditioned reflexes, 41, 69, 72, 74, 83
Forbes, A., and Morison, 68
Folk, G. E., Jr., et al., 24
François, M., 24
Freed, H., *see* Spiegel et al.
Freeman, G. L., and Sharp, 18
Frequency analysis of EEG, 76, 85
von Frisch, K., and Lindauer, 10
Frog, 77
Frolov, Iu. P., 8
Frontal lobes, 25

Galambos, R., 62
Galvanic skin reflex, 42, 45, 82
 and brain stem, 62
 and bulbar reticular formation, 61
 and central nervous system, 62
 and cerebral cortex, 62
 conditioned inhibition of, 61, 69
 conditioning of, 43
 during drowsiness, 61
 and EEG changes, 45, 47, 56
 gradual weakening of, 58, 61
 and K wave (complex), 60, 61
 latency of, 53
 mechanism of variation of intensity of, 61
 recording of, 34
 refractoriness of, 69
 and reticular formation, 62–63
 and skin sensitivity, 59, 61
 as a temporally conditioned reflex, 31, 44
 trace-conditioned, 44–45
 unconditioned, variability of, 43
 See also Skin response
Gambarian, L. S., 8
Gardner, W. A., 24
Gastaut, H., 54, 68
 See also Yoshii et al.
Gellershteĭn, S. G., 2
Gerard, R. W., *see* Libet and Gerard
Gildemeister, M., 58, 61
Gilliland, A. R., *see* Schaefer and Gilliland
Glass, R., 17, 19, 22
Goadby, H. K., *see* Goadby and Goadby
Goadby, K. W., and Goadby, 58, 61
Gooddy, W., 22
Goudriaan, J. C., 23
Grabensberger, W., 24
Granit, R., 62
 and Holmgren, 62
 and Kaada, 62
 et al., 62
Grassmück, A., 21
Green, J., *see* Roth, M., et al.
Grindeland, R. E., *see* Folk et al.
Grossman, F. S., 7
Guilford, J. P., 21
Gulliksen, H., 20
Gurevich, B. Kh., 78

Hagbarth, K.-E., and Höjeberg, 62
Halstead, W. C., et al., 30
 See also Walker et al.
Harton, J. J., 18, 20
Harvey, E. N., *see* Davis, H., et al.; Loomis et al.
Hawickhorst, L., 18
Hernández-Péon, R., et al., 62
 See also Scherrer and Hernández-Péon

INDEX 117

Hibernation, 24
Hill, C. J., and Calvin, 69
Hill, D., and Parr, 29, 76
Hoagland, H., v, 24, 76
Hobart, G. A., see Davis, H., et al.; Loomis et al.
Hoffmann, K., 13
Höjeberg, S., see Hagbarth and Höjeberg
Holmgren, B., see Granit et al.
Holubář, J., vi, vii, 28, 34, 38, 47, 59, 60, 62, 68, 69, 79, 80, 109-11
and Kohlik, 30, 81
Holubář, J. (Sr.), viii
Homack, W., 21
Hormia, A., 20, 21
Hourglass, 23
Hull, C. L., 9, 65
Hülser, C., 20
Hyperventilation, 81
Hypnosis, 5
Hypothalamus, 24-25

Imposed rhythms, electrotonic propagation of, 80-81
See also EEG rhythms, imposed; Flicker
Impulses, nonspecific afferent, 68
Indifferent (optimal) interval, 16-17, 18
Infeld, L., see Einstein and Infeld
Influenza, absence of galvanic skin reflex in, 43
Inhibition
conditioned
of galvanic skin reflex, 69
of unconditioned reflexes, 69
in intervals between temporally conditioned reflexes, 68
Interrupted light, ix
See also Flicker; Photic stimulation
Israeli, N., 22

Jaensch, E. R., and Kretz, 20
Janet, P., 19

Jasper, H. H., vi, 30
and Shagass, v, 9
Job, C., see Granit et al.
Jouvet, M., see Hernández-Péon et al.

K wave and complex, 47, 58
and brain stem, 68
characteristics of, 47
as a conditioned reflex, 66
conditioning of, 47, 68
and diffuse projection system, 68
and galvanic skin reflex, 47-54
dissociation of, 61
mesencephalic origin of, 68
and movement (blink) artifacts, 49
and nociceptive stimuli, 49
and nonspecific afferent system, 68
predominance at vertex, 68
and secondary cortical response, 68
and skin response, dissociation of, 69
and skin sensitivity, 60
and sleep vs. waking state, 53-54
as a temporally conditioned reflex, 67
Kaada, B. R., see Granit and Kaada
Kahnt, O., 17
Kanaev, I. I., 9
Kayser, C., 54
Knox, G. W., see Halstead et al.
Kochigina, A. M., see Dmitriev and Kochigina
Koehnlein, H., 17, 18
Köhler, W., 17
Kohlik, E., see Holubář and Kohlik
Koniukhova, V. A., see Kvasnitskiĭ and Koniukhova
Korsakov's psychosis, 25
Kretz, A., see Jaensch and Kretz

von Kries, J., 20, 21
Krzhishkovskiĭ, K. N., 6
Kuiper, J., *see* Bekkering et al.
Kuthán, V., viii
Kvasnitskiĭ, A. V., and Koniukhova, 8

Labyrinth, of birds and navigation, 12
Larsson, L. E., 68
Laufberger, V., 110
Law of Janet, 19
Leonow, W. A., 45
Libet, B., and Gerard, 77, 81
Light, velocity of, 1
Lindauer, M., *see* von Frisch and Lindauer
Livingston, R. B., 62
Loomis, A. L., et al., 30, 47
See also Davis, H., et al.
Loranz, M., *see* Omwake and Loranz

Maiorov, F. P., 8
Mammillary bodies, 25
Matthews, B. H. C., *see* Adrian and Matthews
Matthews, G. V. T., 10, 12, 13
McClelland, D. C., 18
McCulloch, W. S., *see* Dusser de Barenne and McCulloch
Mehner, M., 17
Meltzer, M. R., *see* Folk et al.
Merton, P. A., *see* Granit et al.
Metronome, 6, 9
Meumann, E., 16, 19, 20
Migratory birds, 10
Morison, B. R., *see* Forbes and Morison
Morison, R. S., *see* Dempsey and Morison
Müller, H. W., *see* Schütz et al.
Münsterberg, H., 21, 22

Navigation of birds, vii, 6, 10–13
and flicker, 85
and local (landmark) characteristics, 12
and local time, 11
random course in, 12
sense of time in, 12
sun arc hypothesis of, 11, 12
and visual acuity, 12
Neurons, rhythmic oscillations of, and alpha rhythm, 77
synchronization of activity of, 77
Nikiforovskiĭ, P. M., 8
Nociceptive stimuli, 34
Nonspecific afferent system, and K wave, 68

"Octaves, rule of", 73, 74, 83
Omwake, K. T., and Loranz, 5
Orchinik, C. W., *see* Spiegel et al.
Orientation of birds, and artificial sun, 13
Orienting reflex, 42
Oscilloscope, 35
Oswald, I., 68

Pacemaker(s), 29
for time, 25
Paillard, J., *see* Bancaud et al.
Pampiglione, M. C., 69
Parr, G., *see* Hill and Parr
Pauli, R., 15
Pavlov, I. P., 9, 22, 64, 65, 74, 82
Pendulum, 5, 22
biological, 22, 27, 41, 74
brain rhythms as a time-measuring, 85
commensurability with time being measured, 26
and metabolism, 23, 24
See also Biological rhythm(s)
Perception(s)
auditory, 2
durations of, 3
of space, 4
tactile, 4
temporal aspects of, 2
visual, 2, 4

Philip, B. R., 18, 22
Photic driving (entrainment) of EEG, vi
  See also EEG rhythms, imposed; Flicker
Photic stimulation, vi
  See also Flicker
Photostimulator, 32, 42
  See also Stroboscope
Pigeon breeder, vii
Pigeons, homing, 10
Pimenov, P. P., 7
Pneumograph, 38, 80
Pruvot, P., *see* Yoshii et al.
"Psychical present", 17
Pulse rate, neural control of, 79

Quasebarth, K., 20

Rabbit, 78-81
Receptor, sensory, 68
Reflexes, *see* Conditioned reflex; Galvanic skin reflex; Orienting reflex; Temporally conditioned reflexes
Refractoriness, of galvanic skin reflex, 69
Respiration, recording of, 37, 38
Respiratory center, 81
Reticular formation, 68
  bulbar, and galvanic skin reflex, 61
  centrifugal effects of, 62
  and galvanic skin reflex, 62-63
  inhibition of, 62
  and skin sensitivity, 62
  temporary connections in, 68
Retina, 4
  cortical projection of, 4
Rhythm(s)
  autonomous, 83
  biological, *see* Biological rhythm(s)
  brain, *see* Brain rhythm(s)
  imposed EEG, *see* EEG rhythms, imposed
  internal, as a pendulum, 5

  production and reproduction of, (tempo), 16, 39
  and exercise, 23
  and heart and respiratory rates, 23
  and sense of time, 31
  specific periodic, 2
  time-measuring, 83
  alpha rhythm as a, 75-76
Rhythmic activity, in isolated cortex, 77
Rhythmic processes, physical, 28
Rocket, 1
Rodnick, E. H., 69
Roth, B., 54
Roth, M., et al., 47, 53, 68
"Rule of octaves", 73, 74, 83

Sänger, E., 1, 2
Schaefer, V. G., and Gilliland, 23
Scherrer, H., and Hernández-Péon, 62
  See also Hernández-Péon et al.
Schönenberg, H., *see* Schütz et al.
Schulz, B., 17, 20, 21
Schumann, F., 20
Schütz, E., et al., 76
Secondary cortical response, and K wave, 68
Sense of time, x
  and alpha rhythm, 76, 78
  and bird navigation, 12
  in birds, 11
  and body temperature, 24, 31
  and brain rhythms, 74
  and cerebral cortex, 25
  definition of, 1
  and flicker, 75
  during sleep, 85
  and electroencephalography, ix
  and frontal lobes, 25
  and heart and respiratory rates, 26
  investigation of, 1
  and Korsakov's psychosis, 25

Sense of time (*continued*)
  and mammillary bodies, 25
  mechanism(s) of, ix, 9
  and autonomous rhythms, 26
  and heart beat, 22
  and respiration, 22
  and rhythmic processes, 22
  See also physiological basis of
  and metabolism, 23
  ontogenesis of, 19, 76
  and other fields, vii
  pacemaker for, 29
  and pharmacological agents, 24, 31
  and physical culture, 2
  physiological basis of, vii
  See also mechanism(s) of
  and posthypnotic suggestion, 5
  and predetermined awakening, 5
  and production and reproduction of a rhythm (tempo), 31
  psychological and physiological aspects of, v
  psychological study of, 14, 82
  motor and sensory methods of, 15
  temperature dependence of, 76
  and temporal lobes, 25
  and temporally conditioned reflexes, 31
  and thalamo-cortical connections, 78
  and thalamus and hypothalamus, 24–25
  and thyroid disease, 24
  and trace-conditioned reflexes, 31
  See also Time, estimation of
Sensory perceptions, temporal aspects of, 2
Servít, Z., 109
Shagass, C., *see* Jasper and Shagass
Sharp, L. H., *see* Freeman and Sharp
Shaw, J., *see* Roth, M., et al.
Simon, T., *see* Binet and Simon
Simonson, E., and Brožek, 17

Skin potentials, 33
  recording of change of, 36–37
  reflex changes of, 63
  See also Galvanic skin reflex
Skin resistance, 33, 42, 62
  measurement of change of, 35
  variation of, 43
  vasomotor origin of absence of, 43
Skin response, 69
  See also Galvanic skin reflex
Skin sensitivity
  control by reticular formation, 62
  and galvanic skin reflex, 61
  and K wave (complex), 60
  mechanism of variation of, 61
Škorpil, V., *see* Drechsler and Škorpil
von Skramlik, E., 3, 15, 17, 18, 21, 22
Solar system, 1
Somatic sensation, cortical projection of, 4
Space
  perception of, 4
  special sense for, 3
Spiegel, E. A., et al., 25
Spindles, EEG
  synchronization of, 79
  with respiration in the rabbit, 79–80
Stein, P., *see* Wang et al.
Sterzinger, O., 24
Stimulus
  conditioned and unconditioned, 42
  indifferent, 67
  nociceptive, 34, 61
  and K waves, 49
  responses to, 59
  unconditioned, 44
Storm van Leeuwen, W., *see* Bekkering et al.
Stott, L. H., 18
Straus, E., 25
Stroboscope, 34, 35
  See also Photostimulator

Stukova, M. M., 7, 69
Sun, artificial
  and bird orientation, 13
  and navigation of birds, 10
  and orientation by pigeons, 12
Sun arc hypothesis of bird navigation, 11, 12

Tarchanoff, J., 31
Television, transmission of an image by, 4
Temperature, body
  and alpha frequency, v
  and EEG (brain) rhythms, 24, 76
  and estimation of time, v, 24
  and sense of time, 24, 76
Tempo, see Rhythm(s), production and reproduction of
Temporal conditioning, vi, 83
  of galvanic skin reflexes, effect of flicker on, vi
  as method for study of sense of time, 15
  See also Temporally conditioned reflexes
Temporal lobes, 25
Temporal sequence of sensations, 2
Temporally conditioned reflex(es), 6–10, 45, 53, 61, 63, 82
  accuracy of intervals of, 7, 8, 9, 44, 64, 83
  anticipatory, 63, 64, 65
  in children, 9
  criteria for establishment of, 63
  defensive movements as, 8
  difficulties in studying, 84
  effect of flicker on, 41, 69–74, 79, 83
  elaboration of, 47
  extinction of, 7
  galvanic skin reflex as an indicator of, 9, 31

  inhibition of, 7, 47, 69
  in the intervals of, 68
  and internal rhythms, 9–10
  K wave as a, 67–68
  method of, 42
  minimum intervals of, 8
  patterned movement as a, 7, 8
  production of, 44
  reinforcement of, 44
  and rhythm of stimulus, 7
  and rhythms, 24-hour, 8
  salivary, 6
  and sense of time, 31
  theories of mechanism of, 9, 64–65, 74
Temporary connections, in reticular formation, 68
Thalamo-cortical connections
  and origin of alpha rhythm, 78
  and sense of time, 78
Thalamus, 78
  and sense of time, 24
Time (intervals)
  aboard a rocket, 1
  autonomous, 1
  biological, 2
  and aging, 23–24
  mechanisms for measurement of, 3
  "center" for, 84
  dimensions of, 2
  estimation (judgement) of, v, 9
    accuracy of, 4
    and "attentive set", 20
    and body temperature, 24
    and dorsomedial thalamotomy, 25
    and emotional factors, 4, 20–21
    and fatigue, 21
    filled vs. empty intervals, 19
    and hypnosis, 21
    individual variability in, 18
    intentional alteration of, 20
    mechanisms of, 10
    motor method of, 15

Time (continued)
  production and reproduction, 16
  overestimation and underestimation, v, 8, 16, 18, 21
  and pharmacological agents, 24
  and preferred tempos, 23
  and proprioception, 21
  sensory method of, 15
  and stimulus intensity, 19–20
  subconscious process of, 4, 5
  See also Sense of time
  internal, shift of, and shift of bird orientation, 13
  local, in bird navigation, 11
  measurement of, intrinsic means for, 2
  mechanism for measuring, 6
  on earth, 1
  pacemakers for, 25
  perception of (the flow of), 1, 82
  in dreams, 82
  theories of, 3
  See also Sense of time
  and physics, 82
  special sense (organ) for, 3
Time-measuring mechanism(s)
  and alpha activity, 78
  of the brain, v, 75
  pulse and respirations as possible, 78
Time-order error, 17–18
Toman, J., 30
Trace-conditioned reflexes
  extinction of, 7
  as method for study of sense of time, 15
  and sense of time, 31
Tresselt, M. E., 18

Unconditioned reflexes, 63, 66
  inhibition of, 69
  See also Conditioned reflexes
Universe, 1

Upper respiratory tract infections, and absence of temporally conditioned reflexes, 60

Vasilenko, F. D., 7
Vatsuro, E. G., 8
Vegetative equilibrium, disorders of, and galvanic skin reflexes and EEG changes, 60, 61
Velasco, M., see Hernández-Péon et al.
Velocity
  biological effects of, 2
  of light, 1
Veraguth, O., 31
Vertex, predominance of K waves at, 68
Vierordt, K., 15, 17
Visual acuity in birds, and navigation, 12
Vondráček, V., 24, 25

Walker, A. E., et al., 30
  See also Halstead et al.
Walsh, E. G., 78
Walter, V. J., and Walter, 30
Walter, W. G., see Walter and Walter
Wang, G. H., et al., 61
Water beetle, 77
Webb, H. M., and Brown, 27
Wiener, N., vi, 110
Wirth, W., 16
Woodrow, H., 3, 17, 18, 20, 21
Woolf, J. I., see Halstead et al.; Walker et al.
Wundt, W., 14, 15, 17, 20, 22
Wycis, H. T., see Spiegel et al.

Yoshii, N., et al., 68
Youtz, R. E. P., 69

Zavadskiĭ, I. V., 7
Zeleniĭ, G. P., 6, 9